THE CITY BUILT AT THE SHINY MOUNTAIN

NEGAUNEE, MICHIGAN
1844 TO 1930

Robert D. Dobson

Dobson Publications
224 Shoreline Drive
Negaunee, Michigan 49866

The City Built
at the
Shiny Mountain

Negaunee, Michigan
1844 to 1930

FIRST EDITION

ISBN 978-0-9747708-9-5

Printed by ABC Printers of Marinette, Wisconsin

Published January, 2010
By
Dobson Publications
224 Shoreline Drive
Negaunee, Michigan 49866

PUBLISHED IN THE UNITED STATES OF AMERICA

The Table of Contents

Introduction

Little did I know when Ethel and I moved to Negaunee six years ago that I would read all of the Iron Heralds from the first one in 1873 through 1929, a period of 57 years and over 2,500 copies. I must say, however, I mostly read only the news pages of each. In addition, I have included additional information from other books and maps so that we might go back all the way to the beginning in 1844. The Negaunee Historical Museum contains most of the original copies of the weekly Iron Herald volumes into the 1960's. The Negaunee City Library has microfilms of the Iron Heralds and a microfilms machine for reading them and making copies. The Museum and Library also have a modern set of two DVD's, containing all the pages of the Microfilms. The Museum has them available for purchase.

In reading the 57 years of the Negaunee newspapers, I recorded 549 pages of typed notes that are available on a CD. The "Dobson Negaunee Iron Herald CD" has a great many more notes than are in this book. Using a computer, the CD can locate any words, numbers, topics, streets, weather, stores, schools, or family names, mines, etc. It is a good supplement to this book in locating something you've read, but cannot later locate it in the book. It also often gives more details than I was able to include in this book.

I have tried to list many Negaunee and Palmer and surrounding town mine deaths in Marquette County. I have recorded the deaths of women who were notable for one reason or another, and I have recorded many of the names of children who died and how they died. It is evident that many families had suffered a loss of at least one member.

I learned a lot in taking these notes, and I am sure you will learn a lot as well. It was fun to relive the history from the beginning and see the good times and the sad times, and new inventions from month to month and year to year. City piped gas brought gas stoves and gas hot water heaters. However, electricity changed everything, and I mean everything. There were hundreds of new electric objects, including the Electric Street Railway, street and store lights, fans, and refrigeration. You will learn about the Cleveland Grove, and Union Park, and all the other entertainment that people enjoyed through the years, including many bands, baseball, girls' sports, and steam boats on Teal Lake.

I would like to say "Thank You" first of all to my wife, Ethel, for her patience while I work in my retirement, and for her proofreading and spelling expertise, finding errors on pages where I have not. I would like to thank Marcia Mattfield and others at the Negaunee Library, Rosemary Michelin at the Longyear Research Library, Roland Koski for his vast knowledge of the history of

Negaunee and its people, and Miles Parkkonen for his creating of the DVDs of the Iron Heralds. I also wish to thank the Negaunee Historical Museum for collecting our city's history, and countless others who preserved the actual newspapers and others who made a record of them on microfilms. I have appreciated the help of the Michigan Iron Industry Museum, and the Peter White Library for its maps. My thanks to Dan Landmark,. Jim Thomas, and Paul Gravedoni. They led me on my first trips through the Jackson Mine and other Negaunee caved-in areas. Thanks to Bill Maki for a tour of an unknown early paved road in his back yard. Thanks to Mike Lempinen and Clifford Stammer for accompanying me on exploratory hikes. Growing up in Ishpeming, it was hard to believe that I had missed so much of Negaunee's local history when I was young.

So start in on Chapter One of this book and enjoy getting carried away into a world of the past.

Robert D. Dobson

P.S.: Those who enjoy this book might be interested also in the following books:

THE EARLY HISTORY OF A MINING TOWN (ISHPEMING, MICHIGAN)
    IT WAS AN UNDERGROUND IRON ORE MINE (SALISBURY MINE)
GROWING UP IN THE SALISBURY LOCATION
    THE PLANK ROAD AND THE FIRST RAILROAD, FROM ISHPEMING AND
    NEGAUNEE TO MARQUETTE
THE RAILROAD THAT NEVER RAN (THE IR&HB RR)
    ISHPEMING'S WILL BRADLEY
ISHPEMING'S J. MAURICE FINN

"THE BRIGHT AND SHINY MOUNTAIN"

Here is a photo taken shortly before this book was published, of some of the remaining "Shiny Mountain" that still can be seen at the old Jackson Mine. This is on the far side of the lookout area and can be seen best on sunny days. We are facing north with the sun behind us. "Shiny" areas can also be seen to the right of the lookout, just a few feet away. This large pit is No. 9 . Pits are marked on the Jackson Mine map inside the back cover.

Bibliography

Adair, Cornelia. My Diary. Austin: University of Texas Press, 1965.

Burt, John S. They Left Their Mark, A Biography of William Austin Burt. California: Landmark Enterprises, 1985.

Boyum, Burton H. The Saga of Iron Mining in Michigan's Upper Peninsula. Marquette: Longyear Research Library, 1977.

Cascade Historical Society. Richmond Township, 1872-1972, Centennial Booklet, 1972.

Cox, Bruce K. Perfectly Safe, The Pabst Mine Disaster of 1926. Wakefield, Agogeebic Press, 2006.

Casey, R. J., and Douglas, W. A. S. Pioneer Railroad (Story of the Chicago and North Western). New York: Whittlesey House, 1948.

Crowell & Murray. The Iron Ores of Lake Superior. Cleveland, Penton Press, 1927.

Cummings, William John. Iron Mountain's Cornish Pumping Engine. Iron Mountain, 1984.

Etelamaki, Levi. Many Shadows, Many Voices. 1998

Etelamaki, Levi. The Blue Collar Aristocracy. Escanaba, Richard's Printing, 1996.

Gwinn Area Historical Committee, Gwinn Area Centennial, 1908 to 2008.

Hatcher, Harlan. A Century of Iron and Men. Indianapolis: Bobbs, Merrill Company, 1950.

Hoffman, Bernard. Reflections from Old Crystal Falls. 1990.

Ishpeming Carnegie Library Calendars. Ishpeming, Globe Printing, 1982-2002.

Ishpeming Rock and Mineral Club. Field Trip and U. P. Gem and Mineral Show. Ishpeming: Globe Printing, 1972.

Ishpeming Historical Society. A Visit to the Past. Vol. I. 2000.

Israel, Wm. H. Souvenir of Negaunee, Michigan. Reprint by the Negaunee Historical Society: Negaunee: Plan B. Publishing, 2007

Ishpeming Historical Society. Ishpeming Sesquicentennial. Ishpeming: Globe Printing, 2004.

Jenkins, Thurston Smith. The Days of Mines (Ishpeming Holmes Mine). Naperville: Bradley Printing Co., 1987.

LaFayette, Kenneth D. Flaming Brands. Marquette: Northern Michigan University Press, 1977.

Lambert, B. J. (Ed.). Baraga County Historical Pageant. Ishpeming: Globe Printing, 1970.

Lankton, L. D. & Hyde, C. K. Old Reliable (Quincy Mining Co.). Hancock: Book Concern Printers, 1982.

Lyman, Barbara. The Mirrored Wall. (Story of Iron Ore in Negaunee-Ishpeming) Ishpeming: Globe Printing, 1973

Michigamme Area Centennial Booklet, 1872-1972. (Ed.) Ishpeming: Globe Printing. 1972.

Whitaker, Joe Russell. Negaunee's Early Years, Dominated by Mining. Negaunee: Plan B Publishing, 2007 (Original Title: Negaunee. Michigan: An Urban Center Dominated by Iron Mining. Chicago: The University of Chicago Libraries, 1931.) Republished by the Negaunee Historical Society, 2007.

Negaunee Iron Herald (Negaunee Weekly Newspaper), Microfilms, Newspapers, and DVD from 1873 through 1929.

Newett, Iron Ore (Ishpeming Weekly Paper). Microfilms 1-15: 1879 – 1920.

Perly, Fred. Upper Peninsula Auto History. Gwinn: Avery Color Studios, 1982

Pyle, Susan N. (ed.). A Most Superior Land. Lansing: Michigan Natural Resources, 1983.

Republic Historical Booklet Committee. 125 Years of History. Marquette, Pride Printing, 1995.

Republic Area Historical Society. A Moving Experience, The Partial Relocation of a Community, Republic, Michigan.

Ronn, Ernie. 52 Steps Underground, The Autobiography of a Miner. Marquette, Center for U.P. Studies, NMU, 2000.

Skelly, John W.. Poems of the Pits. England, Cumbria: Printexpress, 2000.

Stakel, Charles J. Memoirs of Charles J. Stakel. Marquette: Longyear Research Library, 1994.

Stiffler, Donna L. The Iron Riches of Michigan's Upper Peninsula. Lansing: Great Lakes Informant, Series 3, Number 3.

Stone, Frank, B. Philo Marshall Everett. Baltimore: Gateway Press, 1997.

Railroads and Abbreviations

I. M. Railway: The Iron Mountain Railway. A plank road with wood rails covered by metal straps..

I. M. Railroad: The Iron Mountain standard gauge railroad from Marquette to Ishpeming. Now the Soo Line.

C&NW: Chicago and Northwestern Railroad from Chicago up to Calumet and to Marquette at one time.

CM&St.P: The Chicago, Milwaukee, and St. Paul Railroad.

DM&M: Detroit, Mackinaw, and Marquette.

DSS&A: The Duluth, South Shore, and Atlantic Railroad, known also as the "South Shore."

LS&I: The Lake Superior and Ishpeming Railroad was built to haul ore from Iron Mines to Marquette dock

M&O Railroad. It became the MH&O RR, The Marquette, Houghton, and Ontonagon RR, then the DSS&A.

M&W: The Marquette and Western (Mostly Mqt. County area)

M&N RR. Manistique and Northern.

M&SE RR. Marquette and South East.

Major Mining Companies

CCI: The Cleveland Cliffs Iron Co., a merger of the Iron Cliffs Co. and the Cleveland Iron Co., now Cliffs Natural Resources.

J & L: Jones and Laughlin

Oliver: Oliver Mining Co. of Minnesota

U. S. Steel: Part of the U. S. Steel Corporation.

Jackson: The Jackson Mining Company, later part of the CCI.

Schlesinger Mines: (Queen, Mary Charlotte, Prince of Wales, Buffalo, and South Buffalo Mines, or the Queen Group)

Mining Terms Explained

Cage: An open elevator with gates, used for taking men, timbers, and other items in and out of the mine through a shaft.

Caps and Legs: Legs and caps were notched to fit together in crossways in mine crosscuts, drifts and levels to keep the rock roof from falling on miners. Boards were then laid on top of them, the length of the tunnels. Caps and legs were mostly made of wood, but could also be metal in later years. Hard ore mines often did not require caps and legs.

Diamond Drill: This invention allowed miners to go deeper from surface or underground in search of pockets or veins of iron ore. It was a round bit with black diamonds imbedded in it and could cut a round hole from an inch to up to 16 inches in size and formed a core which was removed for study.

Drift: a slightly uphill tunnel built off of a level to get to the ore. More drifts, called crosscuts are often added from drifts in the ore area.

Dry: This is the change house for miners to use and hang up their clean clothes when putting on their mine clothes, and to keep their mine clothes after showering and going home.

Hoist: A large barrel from four to twelve feet in diameter which coils the steel cable on and off of one or two skips. A hoist is also required for the cage.

Hoist House: The hoist house contains the hoist and often steam making equipment to drive the hoists, drills, and pumps before the days of electricity.

Level: The levels are the main drifts from the shafts and are sunk often in fifty to one-hundred foot intervals down in the shaft to begin the process of going to beds of ore. Running a bit uphill, they drain all the water from the mine into the shaft where it is often pumped up to surface from the bottom.

Motor-man: When electricity replaced mules and horses in the mines, the motor-man ran the low, flat, electric engine, to pull ore from the drifts and levels to the shaft and the skips.

Pig Iron: Made in the Upper Peninsula furnaces using a flux. Iron ore was melted to about 3200 degrees with charcoal and impurities separated out, allowing a higher percentage of iron to be sent to open hearth furnaces, and for the separated remains, called "slag" to be left in the Upper Peninsula.

Raise: Tunneling upward in the mine, to get to a level above, or to an ore body higher up, or to remove ore.

Shaft: This is a vertical or inclined hole put down in the ground from a small 8 x 8 foot pattern to 9 x 22 foot, or larger. The shaft could have one or two skips, a cage, a ladder, and place for all wires and piping.

Shaft House: This building usually was over the shaft and might contain a crusher to make ore smaller when it came to surface, and some partitions in which to separate and hold different kinds of ore for future mixing, loading into railroad cars, or placing on stockpiles in the non-shipping winter months. Shafts could also be built simply to get more air into a mine.

Skip: This often was a series of buckets welded together to hold several tons of ore on each trip to surface. When in the shaft house, it was self dumping. In the early days, men rode in and out of the mine in the skip. Skips could have large buckets hung below them to help unwater mines when needed. They often traveled at 1,500 feet a minute or faster.

Stope: A stope is a hole being made when removing iron ore. It can be very, very large in a hard ore mine, some say as big as a city block. In a soft ore, or hematite iron mine, pillars must be left periodically to hold up the roof over the miners. Stopes could later be filled with mine rock from another section of the mine and then the pillars with their iron ore, mined and shipped with other iron ore.

Trammer: Trammers pushed cars to the shafts in the early days, and then back up into the mine workings to be loaded with more ore. They also pushed ore in cars out from the shafts on trestles to dump on piles for winter storage. They were replaced with horses and mules, and then electric locomotives.

Winze: These were openings made between levels some distance from a shaft. They could be used for air circulation, getting to stope ore from above, or for efficiency. They did not go to the mine surface as did a shaft.

The importance of the name of the City of Negaunee is told by Peter White in the article at the right. Above are some of the Native Marquette area Americans and their Wigwam and western clothing from those early years. Taken at Light House Point in 1864. From a postcard at the Negaunee Historical Museum.

Iron Herald 1/21/1898

The following is a communication from Hon. Peter White which will necessarily stand as conclusive, as he is the only person in this neck o'woods qualified to speak upon the subject. Over the signature of Piere Le Blanc he says:

ED. IRON HERALD:—Referring to the item in your last paper about the names of Negaunee and Ishpeming, it is a little more correct to say that Ishpeming means "On High" rather than "*The Highest,*" but that name was fixed upon for that place because Ishpeming was built upon *the highest* land between Lakes Superior and Michigan, as is proved by the fact that its waters run into both lakes. The word Ishpeming means in the Indian, Heaven. It is used to signify "HEAVEN" in their religious books. Negaunee does not signify "Hades," as many insist. Its spelling has been changed a little from the Indian, for instance: Nigana means, I am foremost. Nigani —I take the lead—precede, Nigania —I make him first. Niganis—I am a superior—first or chief. Nigani-wik-weiabikissitchigan means the Chief Corner Stone. The name Negaunee was delved out the forgoing, but Mr. Charles T. Harvey was the man who determined the *spell*. In the Indian the word Ishpeming is spelled "Ish-piming," i e, i in the middle instead of e.

Here is a photo, courtesy of Mark Balzarini, Paul Gravedoni, and Roland Koski, of the west part of Negaunee. It was mostly known as the Jackson Mine Addition, and was present until the 1970's when the Cleveland Cliffs Iron Company was expanding the Mather "B" Mine in this area to the south, all the way to Jackson Park. Today this area, bounded by the black line, is absent of all buildings. It has become known now as "Old Town," and has been reopened after being fenced off for the past 40 years. Some areas are still fenced, however, within this area.

The streets and sidewalks, and the stairs up to some homes are still as they were left after homes were removed. There has been no caving of the land, and so it has been sold back to the city to use as it wishes. You may walk or drive through the area.

1. The Name

Negaunee. It's an Indian Name. The French called the Native Americans the Ojibwa, the English called them the Chippewa. They gave the city of Negaunee its name in 1857. Peter White tells us the meaning in the Negaunee Iron Herald article of January 21, 1898, and as printed on page nine of this book. The choice was the word, NIGANI, meaning "I Take the Lead," or "Pioneer," relating to the city's Pioneer Iron Ore Furnace..

Later, in 1906, Peter White explained that while in the State Legislature, he tried to name the local territory into a township called "Teal Lake," while two other men tried to call it "Pioneer Furnace." Said White, "They wanted me to give them the word 'pioneer' in Indian.....and they were satisfied with it, so adopted it and organized the town." (Letter to Mr. A. P. Johnson, February 10, 1906 --Mgt. Co. Historical Society)

The name of the town did not come at the beginning, however. We are now backing up to 1844, when there was nothing here but some mosquitoes and some surveyors.

1844 TO 1858

2. The "Shiny Mountain"

In his book, They Left Their Mark, John S. Burt tells how his family was involved in surveying the area of Negaunee in 1844, following the ceding of Indian lands to the U. S. Government. In detail, he tells how the surveyors, William Austin Burt and others, arrived at Teal Lake, and how the compass needle never pointed "alike in any two places." They had discovered magnetic iron ore as they surveyed a section line south from a corner at Teal Lake. They were evidently not aware, however, of what the Ojibwa called, the "bright and shiny mountain."

Another early settler here was Philo Marshall Everett. He had heard, like many others, that copper had been found in large quantities in the Keweenaw (Key-when-aw) Peninsula of Michigan by Douglas Houghton. He left his home in Jackson, Michigan, with three others for the Copper Country in 1845. They formed a company called the Jackson Company before leaving. When they got to the Sault, where they needed supplies and help to cross Lake Superior, they met Louis Nolan, a well-known guide. While at the Sault, Nolan told them about what he had seen upriver of the mouth of the Carp (now south Marquette). He had seen rocks which were very smooth and what was more important; they "shined brightly." Mr. Everett showed Nolan some rocks he had, including a copper one, but Mr. Nolan said it wasn't copper he saw.

The four Jackson men forgot all about the Keweenaw Peninsula, and had Nolan bring them to the place where he had seen the "shiny rock" 36 years earlier. The group went up the Carp, but could not find the area. However, they found the Ojibwa Chief, Manjekijik, later written, Marji Gesick, and he directed two of the group to the shiny mountain, which proved not to be gold or silver, but iron ore. Mr. Everett's description of the site was as follows: "There lay the boulders of the trail, made smooth by the atmosphere, "bright and shining," but dark colored, and a perpendicular bluff, fifty feet in height, of pure solid ore." (Frank B. Stone book). The shiny mountain was about 3200 feet westerly of Teal Lake and centered in Burt's section "1". The year was 1845, and thus began,

"The Jackson Iron Mine." The word "bluff" became the word "mountain." It was not only used for that large hill of solid iron ore, but for two other "mountains" to the west. Even the first railroad was called, "The Iron Mountain Railroad."

3. The Forge

A wagon road had to be built between Lake Superior and the mine and while building the road from the lake up, they found a waterfalls on the Carp River and decided to make that the site of the Forge. The heavy iron castings were hauled over this trail through sand and extra large "ditches." A wagon trail was also used to get the iron ore from the Jackson Mine to the Forge. The Forge was unable to get hot enough to melt the ore, but could soften it so it could be pounded with a large hammer, causing some of the rock impurities to fall off like scales. The purified, squarish, iron blooms were then hauled down to the lake in small wagons pulled by teams of six horses according to Sam Mitchell in the Iron Ore, 9/15/1917, issue. Others are quoted as saying the teams of six were mules, or even oxen. The wagon road that went from the forge to the Jackson Mine entered the mine area just west of the former Mather B Mine.

From the Forge to Marquette, the wagon road went up past the Negaunee Township Hall going east, north of the Morgan Furnace, and between Co. road 492 and U.S.-41. There was a large sand problem that caused wheeled bloom wagons, as well as winter wagons on runners to cut into the sand. The Wagon Road was soon rerouted from the area of the Negaunee Township Hall down to Eagle Mills east past the front of the old Morgan Heights Sanitarium, then northeast, to rejoin the original route. It would have joined the original Wagon Road somewhere north of the present Sisters of St. Paul de Chartres on 492. The wagon road

followed Grove Street and is thought to have entered the Lake Superior area on what became, Jackson Street. In 1848, the Forge made ten tons of "blooms." The forge was also called the Carp River Forge after the Jackson Co. leased it in 1850. In the 1870's, there also was a Carp River Furnace in Marquette.

The Jackson Mine officially began in 1847. There must have been more activity at the Forge than at the Mine, as the first U. S. Post Office was at the Forge. The first employees lived in homes there. Nancy Hemminger joined the group of residents there, along with her brother and mother, coming from Ohio. She would marry Mr. George Mall in 1856, just after he had built the "White House" on Teal Lake for his employer, Mr. Reynolds. Mail was addressed to the Forge, care of "Lake Superior." There was little or no mail in the winter.

4. The Jackson Mine Stump

The Jackson Mine was noted by not only a large "shiny mountain" of dark rock, but a large pine tree that grew on it, and from which they measured the distance to Teal Lake. This was told by Mr. Everett to the Government: ""Well, there was a certain pine tree there by the ledge that we marked, and then ran to the southeast corner of Teal Lake, and it was bounded by marking that pine tree the center." In more testimony, Mr. Everett noted that "it was right where we first commenced to mine," and that "it was some ways up beyond where their office (Jackson Mine) is now—to the west." Because of the mining, Captain Merry "had cut it down and sawed it up and made a desk of part of it, and he showed me the desk in the office--." (Pages 42-44, Philo Marshall Everett, by Frank B. Stone.)

It is the guess of the author of this book that the stump was probably moved to a non-ore area, and was seen there for many years, until it burned, and we shall come across that news in this history book.

5. Mining at the Jackson

At the Jackson Mine, the ore was quarried out, according to the continuing testimony of Philo Everett in the book by Frank Stone. That is why so many holes and pits of the Jackson are so rugged. Very little rock was ever removed, or as little as possible, and ore was taken as found, in a large area, then perhaps in a very little area, and then again in a big area. On the other hand, there are smooth pits. They have smooth surfaces because the hole is caused by the sinking of the soil into tunnels and "stope" holes underneath the ground, which the Jackson also has. The pressures of the earth are so great that mine tunnels are constantly squeezed under pressure and the timbers split and bent until they have to be replaced or the tunnel abandoned. Then the ground surface and "roof" of the mine fall in and form a smooth pit.

In 1846 and 1847 the first iron ore was removed from the Jackson Mine. More ore was dug in the next several years and it was treated at the Forge. How much of this ore was treated by the Jackson Forge is unknown. The Forge had men go unpaid, had changed owners, and had "delivery of blooms" problems down to the lake. The route was often steep, and wagon brakes evidently very poor. Sometimes the animals were run over by the very loads they were pulling. As there were no real farms yet, all the animal food also had to be shipped in by boat. There was one major breakdown, when the 18-foot dam made for water power gave way. The Forge stopped production several times before ending in 1854. Ore was then treated at the Marquette Furnace until the Pioneer Furnace was built in Negaunee in 1858. The Jackson mined 10,309 tons in 1858. Then in 1859,

production went to 28,377 tons of iron ore mined, and in 1860, 41,295 tons.

Even before the first "wagon road," a Native American Indian route existed at the "shiny mountain." Years later, in 1908, a Mr. Barney told the Iron Herald that there was a depression "in front of the Merry House," which formed "part of the old trail which swung southwesterly from a point a little to the eastward of Teal Lake and continued," finally reaching L'Anse.

A story appeared in the newspaper on September 11, 1914, regarding wolves at that time around 1855. A couple living in Eagle Mills, Mr. and Mrs. Ernest Haupt, had numerous wolves in the area. On a day when Mr. Haupt had left for Lake Superior for supplies, the wolves overtook the cottage, and on the roof, knocked over the stove pipe. Mrs. Haupt and the children were overcome by smoke in the cabin and it almost seemed they would perish inside, or on the outside. However, she was able to open the door a bit and yet keep it secured, and the wolves eventually left them alone.

6. The Teal Lake "White House"

Miss Hemminger, whom we met at the Forge where her mother worked in 1856, had married George Mall. He worked for a Mr. Reynolds whose "business ventures had prospered," according to the March 15, 1918, issue of the Iron Herald. Mr. Reynolds had built the "White House" on Teal Lake, coming north in the summers to use it as a summer home. They would come with boats, fishing equipment, horses, and carriages, and entertained in "grand style." Mrs. Mall notes that the first summer of their marriage that they lived in a small cottage nearby the White House, and that in winter they were allowed to live in it as caretakers. When the Reynolds ceased to come in the summers, it slowly fell into ruin and was consumed by fire. We shall read about the "White House" from time to time. Mrs. Mall notes that it was located in that little point of land jutting out into the lake across from the Cambria Mine Location. One can still see the square mark of the home's basement.

7. The Iron Mountain Railroad

Mr. Heman Ely showed up at the Lakeshore one day with the idea to build a railroad to the Jackson Mountain, and to the "Cleveland Mountain," and the "Little Mountain." All three sets of owners got together and promoted the railroad. However, Mr. Ely was rather slow. He had the route all surveyed and laid out, but he only built it as he could pay for it. The area from Marquette up to the area of the Sisters of St. Paul was also very rugged and required much digging and filling, all by hand.

The mine owners were anxious as the first U.S. government Lock was being built at the Sault and was to be completed in 1855. It was obvious that the RR would not be completed, so the Jackson Mine and the Cleveland Mine started to build a "plank road, of three inch by eight foot planks laid next to each other to speed up the wagons faster than on the sandy wagon road. To make the job easier, they built it right where Mr. Ely's route was to be. There was a legal battle that the RR won, but the Plank Road was completed in 1857 and called, "The Iron Mountain Railway." It did not work very well, even with metal covered wood rails installed on the planks. Luckily, the railroad was also progressing. As it got towards Negaunee, the Jackson ore was transferred from the Plank to the RR and pulled by locomotive to Marquette. The Iron Mountain RR itself was completed the next year, going to all three "Mountains," in 1858. The three mines became the Jackson, the Cleveland, and the Pittsburg

& Lake Superior. In the meantime, when the Railroad was completed, the Jackson Forge had completed its life. While it operated, other forges had developed to treat ore, and the Collin's had discovered how to heat it hot enough to melt it, and the forge became a furnace. The ore melted, and the non-ore ran off the top as slag, while the purified ore was poured into ingots, called "pig iron." Soon after, also in 1858, a furnace was built in Negaunee, called the Pioneer, meaning "The First," or in Ojibwa: Nigani, "I take the lead." The town had received its name, with a slight change in spelling. .

1858 TO 1873

8. The Pioneer Furnace

Twenty-two men, many with their families, came to the Jackson Mine area in 1857, and built what was called "The Number One Stack of the Pioneer Furnace. In April of 1858, they made their first ingots, just three months after the Collin's. It ran steady until 1860 when it needed a new "refractory lining." In the meantime, in 1859, a "Stack Number Two" was completed. By 1862, 60 men were employed, and as the Civil War was now on, much iron was required. Prices in one year, from 1863 to 1864, went from $45.00 to $75.00 a ton. There was a setback, but short, when the second stack was ruined by fire, but rebuilt in five months in late 1864. The Iron Cliffs Company, which eventually owned several local iron mines in the county, saw the value of the Pioneer Furnace. It then merged with the Pioneer Furnace Company, buying 5,000 shares for $40,000. plus 5,000 acres of wood that could be used for making the required Furnace fuel, charcoal. The year was 1865.

The Iron Cliffs Company, itself had just formed in 1864, with the leadership of Sam Tilden, Wm.

Ogden, and John Foster. Kenneth D. LaFayette, in his book, Flaming Brands, tells us that Mr. Antoine Barabee made the first charcoal for the Pioneer Furnace in a pit near Iron Street.

We also see, now, a view of transportation from Negaunee to Marquette on a nice highway, other than the railroad. It is noted in the building of the Pioneer, that two boilers from a defunct forge, the Marquette, were brought to Negaunee on the Plank Road. It had become the first highway. Where it entered Negaunee, it became "Main Street."

It will be a red town. A Mr. C. A. Dunar visited here and reported in the Herald on October 27, 1881, "The town is built on iron; the streets are rich with iron (fill)....and I believe... that there is iron enough in the immediate vicinity to supply the world for the next 100 years." He adds that, "As a caution to future visitors, I will say that it is not always advisable to lean against or sit upon articles that have been for any length of time, in use, nor for a little boy to slide down the stair banisters, as you will be very liable to carry about with you evidence that you have been in Negaunee."

The Negaunee Iron Herald was one of the first three newspapers in the Upper Peninsula, beginning in 1873. It is at this time we will now begin our look at the city of Negaunee as a town. The Chicago and Northwestern Railroad (C&NW) arrived in 1870 and built a dock to ship ore from Lake Michigan, at Escanaba. The Marquette and Western (M&W) will be built along the Carp River through the present Eagle Mills and behind the present Negaunee Sewage plant on County Road 480, and then over to Silver and Gold Streets. It will have its tracks on the hill south of the present Union Station and cross the Iron Mountain Railroad and Iron Street with a large black cast iron bridge by Captain Merry's home. The Iron Mountain Railroad will

go through several names, including the Bay de Noc, and finally be the Marquette, Houghton and Ontonagon (MH&O). It will then become the Duluth, South Shore, and Atlantic (DSS&A), and when defunct, it became the present hikers', and bikers', Heritage Trail. There will be many more railroads and you will learn about them as we go along.

We now begin Negunee's history in the city's weekly newspaper, The Negaunee Iron Herald.

The Two-Stack Pioneer Furnace above and below. Note the charcoal kilns drawn in 1881 to the east (right) of the furnace below. The location is just south of the Breitung House Hotel. Above: Iron Ore Photo, 7/1/1950. Below: Neg. History Museum.

1873

It is the first issue of the Negaunee Herald Leader. The date is November 13, 1873. The editor and owner was C. G. Griffey. The Iron Cliffs Company is glad to have a town newspaper, and supports it with a large advertisement of their Company Store. The purchased space is rather empty of items, however. The Company says that they are too busy to put any advertising into the ad space. However, they do tell us that they sold 200 barrels of flour last week.

The town has a lot of dogs. The editor says that if there were a dog show, Negaunee would win with the most varieties. A cow was run over by a locomotive this past week, just west of the depot. The first church mentioned is the Methodist Episcopal with the number 225 printed~~~it is the attendance in the Sunday school. A Mr. Ewing has a 50 acre farm two miles south of town on the Cascade Road and has finished harvesting 70 tons of hay and 500 bushels of potatoes. His corn did not mature.

One interesting note in the first paper was that the editor suggests that all the doorways in town should have a number. He must have come from a town with house numbers, and misses them in his new environment. It is going to be a long, long time before Negaunee gets mail delivery and doorways have numbers.

9. The Panic of 1873

Mr. Griffey began his paper at a poor time in the history of this pioneer city. The period has become known as the "Panic of 73." There have been few orders for iron ore. Plus, winter was coming on and much mining was still open pit, and Ishpeming alone expects to have 2,000 layoffs. The Iron Cliffs Company will lay off 300 to 400 men in the area. One man says that he has his coal for the winter and only needs a barrel of flour and another of pork. Because mines are closing for the winter, we see that there are mine horses for sale at $50.00 each. Crime at this time involves the breaking into railroad boxcars and robbing items. People are on the move so often, that each week the Post Office prints the names of people who have unclaimed mail.

The Negaunee mines have strange names at this time ~ the Ada, Allen, Grand Central, Himrod, Green Bay, Spur, Calhoun, Hopkins, and McKenna (on section 6, north and east of town). Others are better known such as the Teal Lake Mine on sections 35 and 36, the Rowland, Rolling Mill, Negaunee Mining Co., and the South Side Jackson, behind and south of present M-28. There, soft hematite ore was being mined with only a pick and shovel and was 44 to 55 percent pure iron.

In Negaunee, there is little U.S. money coming into town. The Jackson Mine is giving its employees their own "four-month paper." The C&NW RR will not accept it, but "greenbacks" are few. People call it "Iron Money." And worst of all, the town's banks have failed. At the Negaunee Bank, and the Bank of Negaunee, people lost almost a million dollars. The first lost $600,000, and the balance in the other. It not only involved the savings of individuals, but the deposits of the businesses in town. Mr. Griffey, writing in the July 3, 1914, newspaper, notes that no one had much relief help, and "The winter was an exceptionally severe one." He continues: "the citizens with true Spartan spirit valiantly struggled through the long period of destitution..."

1874

It is just after Christmas. There is an ad in the paper from a Mary Smith. She is looking for her husband, James Smith. A man and a woman wrapped in a shawl in January were found drunk and lying on Iron Street at 11:00 p.m. The Pioneer Furnace is averaging 48 tons of pig iron daily. Fishermen on Teal Lake are busy eating "fine speckled trout" they caught through the ice." One "three-pounder" was caught.

It has become unlawful to race or run horses on Iron Street. Ladies' dresses this year will contain about 180 square feet of calico. The Jackson Street School, already built, lists its Roll of Honor with the name of all students who have had perfect attendance recently. The list is long. Fresh eggs are 30 cents a dozen, and a local man has invented a protection hat for miners made of cloth and cork. Fire protection will improve with the city buying a new large Babcock fire engine using steam. People are coming to see it.

It is now March, and people are shedding heavy coats and traveling more. Pickpockets are at train stations as trains are coming and going. On the trains, people are told not to be taken in by "three-card montes" who want you to gamble for the correct card of the three. All of the area furnaces seem to be still operating in spite of the economy. On the Cliff's Drive, the Tilden Mine area has its own post office and the Iron Cliffs Furnace on the Drive is in blast, and so is the Morgan Furnace, just north and east of Eagle Mills. The Morgan had a large fire late in the month when 70,000 bushels of charcoal "took fire" and Ishpeming, Marquette, and Negaunee Fire departments all responded with special rail transportation. Negaunee brought its hand pumper. And in Negaunee, Mr. Heyn opened his new store for dry goods.

10. Large Negaunee and Ishpeming Fires.

In Ishpeming, on the date of April 23, 1874, the city of Ishpeming caught on fire. The Negaunee Fire Department, of course, responded to help, but upon reaching the Cleveland Location heard the bad news that the city of Negaunee was also on fire. They immediately turned the horses around, and in the rush, the new Babcock Steamer turned over on its side. Being very heavy with water for steam, and being hot, it was impossible to right. Marquette sent 400 people by train, and about half stopped at Negaunee to help, and the rest went on to Ishpeming. The Iron Herald newspaper building was saved and the heat did not get high enough to melt their lead type. A list of buildings consumed was listed, and most, if not all, were of wood construction. Two weeks later, on May 7, we find that new buildings of stone and brick are being rebuilt in Negaunee.

The fire was bad enough, but many people are leaving because of the economy. The C&NW sold a total of $11,777 of passenger tickets in just the month of April alone. In June, 196 tickets were sold when 100 men left for the "Red River Country." Three men were caught in town for counterfeiting "Jackson iron money." The leader of the group was Mr. Peck, the former town marshal. The stamp used to print the money was found, as well as $16,000 yet uncirculated.

Transporting of ore on the Great Lakes is often now by barges and one steam boat can pull several at a time. The Jackson Mine is building an incline railway at Pit #7, and there are several deaths each month and even more serious injuries. Most seem to be from Jackson Mine blasts gone wrong.

Little Josiah Thomas died when a horse kicked him in the head, and the Mann and Kramer store in town has an ad for human hair pieces. At Winter's Opera Hall, a theatrical group appeared and will become very popular spring and fall. Negaunee has its own trotting park, and it has horse racing. The purse in May is $400. Two big circuses are coming, both in June. Both will have a free street parade of horses, elephants, and other menagerie. Tickets will be $1.00 for adults and only five cents for those under ten. Before one of the circuses arrived, it is reported that they have lost an alligator because of cold weather. The newspaper learned from other newspapers, that the other circus will be a "humbug." A group of men tried to hold a dog fight in town but they could not get the dogs to fight.

This is the time period when buffalo disappeared. The paper reports that hunters by the hundreds are gathering "for a grand slaughter" as the buffalo begin to move in immense numbers west of Fort Dodge.

Not all is bad. The Iron Cliffs still has 800 men employed, and their Pioneer Furnace is producing about 212 to 240 tons of pig iron a week. The total payroll of the Iron Cliffs is about $60,000. a month, and is doing okay during this financial panic. Just a month later, its personal currency is being withdrawn, with men again being paid in government greenbacks. 11 men at the Jackson #5 pit mined 1,700 tons of ore in June. It is hauled by wagon down Iron Street to the Furnace. A new industry has begun which will thrive. It is the Richardson and Shipley Carriage making company. The paper reports that the city population is now 3,741 people, and 79 "milch" cows. And in July, the paper tells us that "the worst of the business depression seems to be past." In fact, the stack no. 2 of the Pioneer furnace has run 13 months steadily, producing

about 18 tons daily. We have the first mention of a miner's strike, but it does not affect the Negaunee area mines. The Detroit Infantry, however, is sent to the Republic and Ishpeming areas. The paper makes a note that wages are low, but it is a poor time to ask for an increase.

Some news from out of town is exciting. Rev. Henry Ward Beecher is having wife and girl friend problems in the news each week. Like a live soap opera. Eventually we hear that he is building a large new home. The editor wants to know what kind of a home he is building on the other side?

Doctor Cochran was out looking for some laborers to hire with his horse and carriage when he ran over a little boy.... only slight injuries. The First National Bank is building at the corner of Silver and Iron Streets. Some residents are moving---all the way back to the old country. The Ed Breitung family lost a six-month old to cholera. And businesses are sporting signs reading, "TERMS CASH." And at Morgan Furnace there is also now a foundry, and it is making iron castings for the county area.

People are finding it interesting that there are several "cold season" store owners on Iron Street. They are Winter, Frost, and Snow. The newspaper notes that they are printing 600 newspapers each week on a #7 hand press. Further down the street, the men blasting at the #7 pit (now fenced in) are getting exuberant in their efforts. A chunk of ore weighing 800 lbs went through the window of the D. G. Stones building and demolished most of the show cases and items inside. Another chunk went into Mr. Breitung's residence.

We come to the first mention of women's suffrage, with a column on their right to vote. The city is in the news as it has decided that it needs a bell to be rung when there is a fire. And in sports, baseball is big. Negaunee placed third this year after Houghton and Ishpeming, in the final tournament. In the fall a Texas group, with their own railroad car, came to town to see the mines for investment. One, Mrs. Adair, kept a diary, and noted that they spend a day looking down into many excavations and up at many furnace smokestacks in the area, and hearing about hematite, magnetic ore, and unusable jasper ore. She heard claims that even in the present "bad times," some mines are making 10% profits, but the roads in Negaunee are bad. She says "I suggested 5% for one year and a good road, but they did not see it." She also noted that many new mines seem to have been recently discovered, and that they were all for sale to them at the same price, $100,000.

The U. S. Treasury men also came to town in October. They wanted to check on this "Iron Money" currency they had heard about. In the meantime, the people in town who are managing are reminded that cold weather is coming, those around who are needy should be remembered, especially "married men with families who have no prospect of work this winter." And a very sad accident happened when a little girl, picking up coal from the tracks had her head severed by a backing up MH&O train. The ore season ended this year with almost 250,000 less tons of ore being shipped than in 1873.

At the north Jackson, a piece of ore fell out of the bucket (no skips, yet.) and hit a miner on the head, breaking his skull. He lived, but a Mr. Lander was killed at Jackson Pit #4. He was going up a too perpendicular ladder and it went backwards into the pit below. The Iron Cliffs will keep about 800 men working and has money to pay them. However, a week later it reduced all wages at mines and Pioneer Furnace by ten percent. The Rolling Mill Mine will hire more men, if possible. Many have left. The C&NW has had to raise wages from $1.00 to $1.25 for a cord of wood (burned yet in engines), and notes that "labor is still hard to find." The railroad also notes that there has been a "stampede of men from the iron district."

Mr. McKenna has opened a new restaurant next to his hotel on Silver St.. There was also this year, a grand reopening of the St. James Restaurant. Iron Cliffs paid its employees their monthly wages in November of $20,000. On Thanksgiving Day, the Methodist Episcopal Church had a service at ten a.m. Sheply and Co. are new city manufacturers of carriages. Mr. Sam Collins has opened a meat market on west Iron, and editor Griffey notes that this is the eleventh meat market in town. Mr. Collins will sell a side of beef for seven cents a pound and a whole hog for ten cents a pound. A man ran a team of horses down Iron St. and got 60 days in the county jail. The Morgan Iron Co. has closed its store after selling all the contents. The Negaunee Library is filled with patrons on Friday nights. The Jackson Pit #7 has been closed for the winter, and there is some pressure in the U. P. newspapers for a separation of the U. P. as a separate state.

The large Alexander Maitland home built in 1875 at Main and Healy Streets. Removed for Mine Caving after 1930. Photo by Wm. H. Israel in 1912, the Negaunee Historical Museum, and Harlow's Wooden Man, spring, 1977.

1875

A Charles Howard who killed a man at the Negaunee Bon Ton Saloon in 1873 has now been arrested for a murder in Iowa. At a horse race on Ishpeming's Lake Angeline, it was noted that four buffalo robes were stolen from a carriage. At the Jackson #4 pit, the rope to signal the hoist man broke and a Mr. Fitch had a wild skip ride. It is now January, and men and families are arriving here, looking for work. The paper notes that there is hardly enough for the poor here now.

A large "hoopla" occurs over the building of a railroad called the Marquette and Mackinaw. It was planned before the "Panic of 73" occurred, and now there is talk of it continuing into existence. The government is thinking of taking back the land grant. People are writing long letters for and against it. It will go on all year. In the meantime, all is not well. Mr. Ed Breitung, president of the Michigan Iron Co., sees it go into receivership. Wood and hay are being stolen this winter. Some arrests. A large blinding snowstorm last week blew the fireman out of the locomotive cab of a train coming to Negaunee at the Pioneer Mine, and he wasn't missed until the train got to the station.

Two people were arrested together for indecent exposure in January, Mr. August Van Dorp and a Mrs. Wales. Mr. John Johnson, a former saloon owner here, may hang in Ohio for a murder. He says he is innocent, and citizens of Negaunee are writing letters stating that he was a fine citizen while living here. At the Cliffs Store, seven full grown rats were recently caught in a large bag with flour in the bottom. The results of the U. S. Revenue men being here is that the U. S. government will now begin charging money for handling the "iron" money.

11. Cold Weather

The paper for the first time has talked about weather. It is now March 4th, and records show that only three mornings have been warmer than three below zero since December 18th. For six weeks, the average morning temperature has been 19 below zero. We should add that the ice on Teal Lake is measuring 32 inches compared to 24 for last winter. It is being cut for summer. In spite of the temperature, Mr. Neely has decided to build a new two-story hardware store next to the Bon Ton Saloon.

12. The Jackson School

The Jackson School is already operating, and to think my wife and I went to dances there in high school. Someone wrote a nice poem about the school and it is printed in the paper. People are still moving away. The C&NW has sold 6 tickets to Salt Lake City, 3 to Portland, Maine, and several others. Two deep Negaunee wells have frozen. One was 40 feet deep and one 35 feet. We also know that there is a city hand-powered fire engine. It operates with a handle or two. We find that a group of Negaunee citizens have formed into a fire company and a week later have set up a code of horn blasts to send messages to the volunteers and to the city. If you ever read a book about Virginia City, Nevada, you might read about Jim Chatham, formerly a Iron Cliffs employee, as he was one of the two fighters in a prize fight there. A large, cold snowstorm about March 15 stopped all trains. The C&NW was stuck at Goose Lake, and the MH&O had several freights and passenger trains, some with plows and two engines, unable to move. Some people had frozen potatoes and it was suggested that if you place them in cold water for six days, they will be like new. Lake Superior is so frozen that a stage crossed from Grand Island to Marquette on the ice the week of April 22, 1875. Even on June 17th, the Iron Herald reported that there was a heavy frost and ice a half-inch thick. A hail storm the same week "the size of eggs" broke many windows around town.

The Jackson Mill burned at a cost to the company of $15,000. As is usual, rebuilding begins immediately and a L'Anse brickyard will supply 100,000 bricks for the new mill and machine shop to be built. About children: Elrick Crimpeau of Negaunee plays marbles with friends who do not cheat. Elrick, who is 12, weights 107 pounds. At Winter's Hall, there was a nice spelling tournament with nice prizes. And the deadly typhoid has showed its head with two cases. The usual special rules have been put into effect. Typhoid will affect many children in coming years.

It is spring and more railroad crime. The MH&O has two satchel thieves riding the rails with passengers, and the C&NW has the card-playing thieves once again. We find out that Ohio will not hang Mr. Johnson, but give him a life sentence. The bank here says that he has money in an account that belonged to the man he murdered and doesn't know what to do with it. Another local man tried to commit suicide by laying his head on the tracks. The oncoming train was equipped with new Westinghouse brakes and stopped before reaching him. Native Americans have been in the Republic (Iron City) area selling their skins (hides), but have now gone to Marquette, hoping for better prices. A new bowling alley has been opened by Mr. Engles and he will keep down rowdyism. A local hotel has also been redecorated and renamed by Mr. Marcotte. He may be Canadian, as he proudly renamed it "The Montreal."

13. Things Improve During the Summer

It is now almost June. Mines are reopening. The Teal Lake is working with 40 men, and the Himrod, Goodrich, and McComber are getting going. The Jackson will pump out pit #7 for mining. The Pioneer Furnace is running both stacks. People are getting ready to travel more and so there will be a passenger coach to Republic twice a day. Often the cars may be part of an ore or freight train. And if you are a coin collector, you will note that the government is now producing a twenty cent coin, a little larger than a nickel. The temperance movement is also growing, but slowly. The Ogden House will now become a Temperance Hotel with no bar or liquor. A Mr. Greenfield has opened a shoe and boot manufacturing shop. Wolves are still very common, with a hundred pound one being killed near Little Lake.

It was discovered after the Pit #7 was pumped dry that the community "Merry Spring" nearby had gone dry as well. A former resident writes from Silver City, Idaho, of the trip there in 12 days, and tells how all the other former residents there are. The Clifford's Dramatic Co. and Sadler's Great English Circus will be here to perform. Reports in other papers say that the show is excellent. Cliffords was so well liked, that it returned here for two more shows after leaving the Copper Country.

14. The Maitland Home

Housing in town is very tight this summer, and only a few homes are being built. One is Mr. Maitland's, at Main Street and Healy. It will be so large that when homes were moved from that area because of caving ground, it had to be torn down.

The White League, working at night, went to Kirkwood's Drugstore and removed the cigars from the statue's hand and placed a small suitcase there. A larger problem will be that "Paris Green," a poison, has been okayed for killing potato bugs. It will be in every home and there will be countless deaths from it, both accidentally, and suicide. There has been a "Cardiff Giant" on display here in town, but it has gone south to another town on the C&NW. Baseball is a popular sport. Negaunee got off to a tough start and lost several games, the first being to Marquette, 28 to 11. In the city limits a band of gypsies have been camping but were soon routed. A black pianist, "Blind Tom," came to town, and all who played the piano were invited to play, and he would assist with an accompaniment, or second part. Doctors have concluded that at least half of the crime is caused by periodicals. A road will be built from Champion to Michigamme. (The original road went around the lake on the south side.)

Now, back to children for a bit. One ten-year old who has a bad habit of stealing has been sent down to the State Reform School. Some other kids are hiding across from the Jackson #7 pit and throwing rocks at passers-by. Captain Johnson ran over a little girl. Both were very frightened, but no injuries. The baseball team played Ishpeming and lost again by 11 to 9. Local mines are paying $1.50 a day. Hematite ore is selling better for furnace use, rather than hard ore, but has less iron content. Women's dresses this spring are so long and tight they can hardly sit down. And the men who formed the fire department are now getting pilings put down for the new fire hall. It will be on the small triangle on the west side of the present city hall.

There is a County Fair, and it is in Ishpeming, and is supported by the County with $500. At this time the county road from Marquette comes from the Marquette Quarry and Poor House area towards Eagle Mills. It is almost impassible stated the paper. By September, the paper notes that this has been a very cold summer. Mary Egan was run over by a train and lost a hand and a foot. The jeweler, Mr. Charles Sundberg, has hired a watchmaker who formerly worked for Tiffanys in New York. The McComber Mine had a setback common to many mines. 3,000 tons of rock fell off the side of the pit, covering the ore. The very next week, 3,000 tons of rock fell into pit #2 of the Jackson Mine. It all had to be removed.

There will be a very popular group of black minstrels who will come to Negaunee quite regularly and are very popular among the town. They came in September and "gave the best show ever seen at Winter's Hall." In spite of the economy, Mitchell and St. Clair plan to open a dry goods store. In spite of the cold summer, potatoes survived okay. Mr. Blake of the McKenna farm planted 5.5 bushels of potatoes and reaped 150 bushels. And a strange action by the paper took place when, after encouraging people to buy locally, recommended that people shop at Jackson's Store in Marquette. Since many people are from Europe and used to eating seafood, a new restaurant has opened, called the Ocean Oyster House. The cold continues and by October 14, sleighs are appearing on the streets. Quite a few people have their potatoes frozen in the ground. The city will dig down a trench to drain the Partridge Creek swampy area through town, especially south end of Jackson Street. The Methodist Episcopal Church will be renting its pews on Friday. High priced ones are in front. Another blast from Pit #7 sent a rock into the head of little Bella Best, age 6, who was sitting right next to the Iron Herald Office. Her brain was exposed, but Dr. Cyr says she is doing okay. Maybe things are looking up. The # 7 pit may operate all winter, and the MH&O wants to buy 5,000 RR ties.

The Episcopalians now have their own church along with a fine $3,000 organ. If you recall the "White House," it is being fixed up by the Teal Lake Mine for its workers. In December, Christmas tree forests are in demand, and Negaunee claims the oldest person in Michigan with Mrs. Belhumeur, born in Quebec in 1770 and has a daughter age 79. She recently walked to Ishpeming with her daughter and still does not need glasses. As usual many mines have suspended operations for the winter.

The Reynold's "White House" on Teal Lake's south shore. It burned down in 1879. This painting of it is at the Negaunee Public Library. The photo below shows where it was along US-41, west of Negaunee.

1876

In January we hear about a raffle for the first time. They will become very, very popular. This time it is called a Grand Raffle, and will include a $170 organ, a sofa, and kitchen furniture. There is no water system yet, and no fire hydrants, so the fire engine hoses must reach from the source of water to the fire. Residents on Cyr Street are asking the fire department to buy more hose. There are several companies in the county that make explosives for the mines. The American Powder company in Negaunee has failed to pay its taxes and the city seized 37 kegs of powder. Several town businessmen have sold out and moved to Colorado. Mr. Reed is selling his house in the Jackson Location with raffle tickets of $1.00 each. The Jackson Company is auctioning off several homes on their property for taxes. They sold for between $120 and $435.

Horses are getting very ill from a disease called "Epozootic." Owners are uneasy. The Morgan Company is going further and further for charcoal. They have 30 kilns on Sections 9, 15, 17, 32, and 35, and 12 kilns near Champion. The Pioneer Furnace reports that they produced 17,000 tons of pig iron in 1875. A train carried 2.5 tons of nitroglycerin from Negaunee to Republic. Captain Dunn has invented a new "Thunderbolt" explosive. It was tried at the Pioneer pit of the Jackson and it worked very well and was safe to use. There was an unscheduled explosion when 38 dogs clashed in a dog fight on Iron Street. The Morgan furnace will go into blast this summer and Mr. C. Donkersley will run it. The Morgan school is running and attendance has been very good. The county road this year will change from going to the quarries in Marquette, to going on the old Plank Road to the Forestville Road. A new larger canal is being added at the Sault for boats. The fire department has bought 1000 more feet of additional hose for $1,500. Republic has a drama association and it will do a production at the Republic Opera House. At Michigamme, there is a beautiful new Babcock Fire Engine, as well as a hook and ladder truck.

Things continue to recover from 1873. The Iron Herald notes on March 30, 1876, that "there are very few idle persons in Negaunee currently." Housing for those seeking it is hard to find. More details in the same paper tell us that the new county road will leave Marquette by Superior and Grove Streets and pass Harlow's old mill, go past the township road to Franklin follow the old MH&O track halfway to Bruce, then go northwesterly to the present state road two miles east of Morgan, and then into Negaunee. The newspaper notes that the leaving of C. C. Eddy and Frank Eddy and their families to Sunshine, Colorado, means that "two, fine, first-class families are leaving our area." A group of youngsters gave a saloonkeeper trouble last week and two had to spend two days in jail. And about that Marquette and Mackinaw RR ~ it plans to begin work this spring.

There was a mine called the Cascade and it was at the Cascade Location. However, the post office was called "Palmer." The newspaper says that it is the name of the person who owned the property there. We find that the Cardiff Giant is owned by Mr. Coburn who formerly owned the town's Ogden House. His exhibiting of the Giant has met with success downstate. Pat Flynn, the town bad guy showed his brutality at Lobb's barroom this week. Several brutes have also appeared from Wisconsin. Said the paper, "Give them a wide berth—normal crooks would not want to be seen with them." Police arrested the 3-monte rogues and shipped them to Green Bay. The newspaper received a warning from towns down the line that "Notorious Canada Bill" would be passing through this way. People are still on the move as well. Six Swedish families from Ishpeming are moving to Minnesota where they have purchased farms, and several French families have returned to Canada. 40 Finns and Swedes came through town from Houghton on the train, on their way to the Black Hills. In the next paper, it is reported that some of the 40 have returned already. Many of those who left in 1873 have returned here, or have written to say they will be coming back. Mr. Maas, Rowland, Mitchell, and Lonstorf have purchased a silver mine in Colorado. The Negaunee steam fire engine could not get up steam, but has been repaired. Here is the list now of the new brick buildings since the fire: Cyr, Laughlin, Mulvey, Quinn and Ryan, McKenna, Mitchell, Kirkwood, First National Bank, and Engles.

Private industries can get pretty big, even in 1876. John Blair is making maple syrup and has tapped 1,600 trees on section 16. Mrs. W. P. Harris is opening up her private school on Main Street. At the Morgan, several families are destitute and are going door-to-door to obtain bread. At the mines, a new blasting powder called "Thunderbolt" has been introduced. The vertical shaft at the Pit #7 will have a new skip and be the main hoist. It is 1000 feet east of the engine house and near the railroad track. Of the 18 furnace stacks in the area, only 6 are currently in blast. Canada Bill arrived in town and the marshal spoke with him and he agreed to leave in 30 minutes. For culture, the Clifford Dramatic group is back in May, and will perform, "Nobodies Daughter." The steamer, "Shoo Fly," is busy on Teal Lake. Several sail boats are also seen there.

Quite a few cows and calves are being bought and sold in town this spring. The fire department is building a reservoir north of the Cyr Street Store at Jackson St. for water needed in case of fire. The old hand pumper has been put into the hands of Captain Merry and it will be rebuilt. June 1, and there are still large ice fields on Lake Superior. J. Meager and D. Cassin received wishes from many friends and have left for settlement in Colorado. The miners of the Lake Superior area use 300 tons of candles every year. The Culex family has been forcing their disagreeable music on the community this week. There was a fine turnout for Prof. Fohrman's concert by the Negaunee Musical Association. The Lent and French Circus and Menagerie show came. People should have watched their clothes lines. The usual amount of stealing took place during the show. Tom McKenna says he will make it lively for the person milking his cow in the morning before he does. 169 boys were born last year and 165 girls. 34 males died, and 53 females. Note the following from the June 22, 1876, paper: The temperature was 35 degrees and it snowed heavily with an inch accumulating at the Iron Cliffs location. Dennis Munn, of the house of ill repute on the old road to Eagle Mills has been sent to prison for a year. Families at Morgan have left, with three finding work here. All vacant homes seem to be filled again. All the snow and heavy rains have caused curtailment at several pits and mines because of the immense water problems. Some potatoes are rotting in the ground, but the hay crop will be very large. Silver has been found 3 miles Northwest of the old Holyoke property, in the Dead River area. A big July 4th celebration is coming and there will be many monetary prizes.

A very interesting note is made of cemeteries about a mile from Negaunee on the Morgan Road in need of care and a note that there are some beautiful monuments and headstones there. Capt. Merry has started a fire company at the Jackson Mine with 100 men and the old city hand-pumper. Note is made that there are four passenger trains in each direction from Marquette to Negaunee currently. Louis Corbett of the Iron Cliffs Co. is leaving for Ohio.

He has been left $20,000 if he moves back. The July 4th celebration is over, and some boys may be arrested for tearing up the Christmas trees put on the streets for the celebration. In comparison with 1873, the year of the panic, there are now not as many mines, and wages are lower. But, notes the paper, people are working and there are no cases of want.

Women's skirts are being made tighter. Workers at the McComber have received their five months of back wages owed them. Scamps are milking people's cows while they are out in fields grazing. The Detroit Opera House Orchestra will be at Winter's Hall for a concert. Michigamme also has an Opera House. On July 21, a heavy frost was seen at the Cascade Location. Negaunee has a 13-year old who gets everything she wants with stories she dreams up. She must be the best liar around and may be the youngest ever in the House of Correction. The girls in town gave a leap-year party for the boys in town. Many guns are in use yet. Wm. Gray had a bullet that went through his hat about an inch from his head while on his way from Ishpeming to Negaunee.

It turns out that potatoes this year are excellent after all and are selling for only 65 cents a bushel from a wagon. A nickel deposit found at Gap, Pennsylvania, is the only nickel mine worked in America. We see that many people can travel great distances now on vacations by train. We see many going to Detroit, Cleveland, Philadelphia, etc. There are 962 children in the Negaunee School census between age five and 20. In late August there is a shortage of working men here. Michigamme has had to recruit out of the area to find 50 laborers.

Mr. S. F. Gilmore is looking for his wife and five-year old son last seen at the Jackson House a month ago. The Cornish wrestling champion is coming to this area and advertises for a challenger. Purses of $100 to $1000. A real surprise was the closing of the Pioneer Furnace. All laid off from the two stacks, and the four charcoal kilns. Men will find other work. Three men in command have left for homes out east. The city is installing cisterns to hold water in case of fires: 12 feet deep and 15 feet in diameter with heavy timber on top. Four flatcars came to town for the nitroglycerin works, a dangerous load. Tom Carkeek of Michigamme has decided to wrestle Mr. Pollard, the Cornish Champion. Sheply and Company are busy making 25 cutters for winter use. The Morgan Furnace is receiving two carloads of charcoal every day from the Champion area. The Negaunee Baseball Team has disbanded and declared themselves the Champions as no other teams have challenged them. Books for school this year will be about $4.00. The Jackson Co. has begun work again at the North Pit, just east of the C&NW track at the tunnel.

There are several bad cases of typhoid now and several fatalities. A large schooner, with a load of coal, and named the "Negaunee," may have been lost in Lake Michigan. It was quite a snow storm for it only being early October, with eight inches of snow and down to 18 degrees. Sleighs are in use in some places. The nitroglycerin plant on the Cliff's Drive will move to Negaunee by Teal Lake. The Jackson is putting down a 200 foot shaft in the pit west of pit #7 to connect to the #7 shaft. The price of kerosene is going up and some may have to go back to candles for light. A little village is appearing at the Cambria and Bessemer Mines.

The Rolling Mill furnace, in operation for over two years, is producing about 1000 tons a month of pig iron for use in Bessemer blast furnaces. It is election time, and the Iron Herald is a strong Republican paper, so it is filled with ads and articles for that party. Most of the mines close for Election Day. The main election result was that Ed Breitung of Negaunee has beaten Peter White for a seat in the Michigan Legislature. The Breitung family promptly left for a vacation in Boulder, Colorado, until he goes to Lansing to serve. Another good thing about Negaunee, notes the newspaper, is that it has two city bands, and "neither has to be ashamed of." Things are now looking so well as Christmas approaches that the editor "believes that another season will show that the prosperity of this region will attain heights never before seen." A fine monument of Italian marble has been received for the grave of Mrs. A. J. Sterling. Sam Collins will have a steer weighing 2,800 pounds on display at his market.

The Breitung House Hotel was built in 1879 and burned to the ground in 1988, over a hundred years later, and still being in use. It was located where the Concert Band building now is. Photo from Harlow's Wooden Man, Spring, 1977.

Below is Iron Street after an 1885 snowstorm. The Pit #7 shaft of the Jackson Mine near Iron Street is in the center background. Negaunee Historical Museum 1988 Calendar.

1877

Mr. Bernard Carr had a thriving candle-making business for all those thousands of men in the Marquette area iron mines. In the first week of the year it burned down. Part of the reason was the poor fire department equipment. They had the new bell, but no rope for it yet, and the steamer, when put to work, had several problems. Captain Merry has come up with a new invention that will allow sleighs carrying ore to be able to dump. "It should be patented." Ore shipments to Marquette totaled about 458,000 tons last year, to Escanaba, 406,000 tons, and to L'Anse 89,000 tons.

15. Scarlet Fever

The prevalence of Scarlet Fever in the county is becoming alarming. The Kirkwoods lost a two-year old daughter. Dr. Cyr has a few cases of it. His recommendation is to keep children warm and keep them from getting colds. A Chicago firm is buying two ores from the Rolling Mill Mine, an ochre, and a bluish, and will use them to color mineral paint. They will use over 500 tons this year. Mr. Heyn's store has gone bankrupt here. In the middle of February the snow disappeared so much that people had to return to using wheels instead of runners. In Lake Michigan, ten million spawn put in included salmon. It is time for Lent, and since there is not too much free money to use, "no one can do anything that requires repentance." A young man in Illinois escorted an older woman home who had only one eye, and no teeth. She died in only a few days and left him $4,000. Notes the editor: "Let some of our youth here try that." Mr. Merry's new steam whistle at the Jackson sounded a fire alarm this week and the fire department saved the building with only $200. of damage.

Store owners, notes the newspaper in April, seem more confident. Also, paper currency is being replaced again by silver. Still, a large group of men left yesterday to go to the Black Hills. A week later, two men came from the Black Hills and reported that there are long lines of unemployed there.

16. Unusual Weather

It never did snow much after February, and by the middle of April it had mostly disappeared altogether. The ground was thawed and Mr. Sam Robbins, had a well collapse on him while he was 16 feet down in it. He was lucky. Mayor Kirkwood has a gymnastic cow. It climbed a set of stairs downtown, went across a stage and on to a roof of an adjacent building. The mayor has written to Mr. Barnum to see if he is interested in her.

Typhoid continues its sweep into the area with five more deaths in the past few weeks. Two were children in one family. There are now 15 "serious" cases. In May there are some new cases again, but it is noted that they are "mild." The Excelsior Peat Furnace in Ishpeming at Lake Angeline has sold its peat drying racks and they will be used at Teal Lake where there is a peat bed nine feet thick on the south shore. The businessmen of town have always been unhappy with stores that come to town, open up, sell their wares, and leave. So new rules have been put into effect for the town and such businesses will now have to pay a deposit of $50.00, so if they leave, they will still have paid their share of the taxes. The Cliff Co. paid out over $12,000. to its employees this week in monthly pay. On May 3, the winter snow arrived--- a little late. Sleighs are out again, and the banks on Iron Street are the highest ever for the winter. And more counterfeit coins. Now they are half-dollars made of glass but silvered over. Crime continues with "urchins" breaking windows in homes that become vacated. Mr. L. A. Marsell had several fine plants taken from his yard and says that the next time they come they may carry away something else as well. A horde of professional gamblers have found a home here. Be careful. There are also pickpockets among them.

William Steehor, 16, died from scarlet fever in a matter of a few hours. On May 31, the son of John Curran, a seven-year old, died from scarlet fever. On June 7th, the son of R. Jackson on Case St., and a Dowing boy at Morgan both died of the fever. They are ages five and seven.

Negaunee has a new baseball team this year and it played Republic in a first game. They lost 27 to 11. Said the paper, "Not bad for the first game." Some people went fishing down on the Escanaba. They brought back 470 trout. And here is information about the fenced in shaft on Silver Street: "McComber has sunk a 28 foot shaft about 160 feet SE of the MH&O Depot. (June 7, 1877). On the week of June 7th, there was more snow in Negaunee. "Well defined flakes as large as goose feathers." Ore vessels arriving in Marquette reported heavy snow between Whitefish and here, with three inches on the deck. An old resident reports that on June 10, 1851, there was 5-6 inches of snow on the ground.

The Georgia Minstrels returned and brought out the biggest crowd since the newspaper went in business. The railroad track from Eagle Mills to Morgan Furnace has been taken up. There will be a big July 4th Celebration at the Negaunee Driving Park, northeast of town. A horse race was held there the week before with two horses. The winner of 2 of 3 races got $300. It is shown on maps in this book. Stop in at Atwaters and see the new furniture and coffins that have arrived. A sidewalk has been provided from town up to the Roman Catholic Cemetery and

will go to the Protestant also as soon as they can cross the Trestle. (This is now in a caved-in area.)

The Pioneer Furnace has burned with a heavy loss. Both stacks have survived, however. We know there is float copper here, dragged down from the Keweenaw area by glaciers, and a 22 pound one was found while stripping dirt off ore at the South Jackson. July 7th was set by the Adventists for the end of all worldly pleasures. It is now July 12. A remarkably good circus came. The Broadway Theater came to town with music, etc. It drew a crowd. A six oar rowboat has appeared on Teal Lake. Raspberries have made an appearance on July 26th, and peaches are also now in town. Eleven new teachers, all misses, yet. And the people of Champion and Republic are hoping for a wagon road between the two towns. Real estate prices have fallen and have not seemed yet to reach bottom. Still a cold summer, as early risers saw a frost on the week of August 2nd. And the worst news of all is that mines are receiving word that they will not open as the price for ore is too low.

Another child has died of scarlet fever, the son of Dan Laughlin. The paper notes that some homes where it strikes are very clean and sanitary, "contrary to ideas of homes being in need of more cleanliness." The State Board of Health requires closed caskets for those dying from the fever, and only private funerals.

Business is not good and Mr. C. Kierens began advertising on May 9th that he will sell all at cost for the next 30 days. The ad is still going and it is about to be September. A large number of men have been leaving the county to work for the Canadian Railroad. The Negaunee baseball team is improving and only lost to Ishpeming by a score of 10 to 4. James Ganzy was out hunting and a boy near by hid in some bushes and made some sounds. Mr. Ganzy thought he saw a crow and shot. The boy, John Heisler, was lucky, and is okay. Thomas Currah was badly injured in a mine accident and was bedfast for two years. Now his wife is also ill and they want to move to a warmer climate with winter coming on. They will sell all their goods by raffle tickets. Negaunee beat Ishpeming in the 4th game, after playing for two hours and 43 minutes. They also beat Ishpeming in the fifth game, 24 to 7. To help the steam fire truck to keep up steam in cold winter weather, the fire department will buy a steam heater to keep the steam up at the station at all times. It will work well. Several heavy frosts in September have killed potato vines.

News reached town that Mr. Maas and Capt. Mitchell were on the Union Pacific train robbed in Nebraska a few days ago with $65,000 taken, and another $1,300 from passengers. Luckily, they were both asleep in a Pullman car that was locked on both ends. The Pioneer Furnace is busy rebuilding from the fire and is now using slate roofs and a lot of metal to prevent another disastrous fire. A local girl answered an ad in a magazine as to how to make an impression. She got a one sentence answer for her money: "Sit on a batch of dough." Negaunee was invited in October to play three games of baseball with Green Bay. The first two games were tied, 13-13, and 11-11. The third game went to Green Bay, but there are hard feelings about the umpiring. Brigham Young was recently arrested in Salt Lake City. The deputy U. S. Marshall was the former Dr. Smith who practiced here in Negaunee from 1864 to 1870. A popular piano song now is "Pretty Little Blue Eyed Stranger." In October, the weather is still bad. The paper editor, Mr. Griffey, reports that the Lion and the Lamb have lain down together. The Lamb is inside the Lion. The Catholics held a fair and raised $1,000 with $700 being profit. A large train-load of new mining equipment came through

town for the Champion Mine. Twenty two railroad cars included six, 7-foot hoisting drums. The new Pendill shaft between Silver and Gold Streets is now down 55 feet. The sale of pig iron seems to have picked up.

Temperance Reform is becoming a big thing and an effort here to get backing brought 248 signatures. It "passed all expectations." Names were printed in the paper. Ishpeming and Negaunee will build a sidewalk between the two towns. The first Thanksgiving Turkey Match will take place. 200 Turkeys have been ordered, and some other birds as well. Someone is breaking the stained glass windows at the Episcopal Church, including some chancel glass that will be hard to match. A meeting was held for raising poultry and an association was formed. Mrs. Harris is closing up her "Select School" and moving her family to California. Two fine citizens, Frank Boone, and Frank Miller have been arrested for robbing Dr. Cyr's office of gold and money. Bail set for each of $2,000. In December, Miss Mary Renan will open up a Select School for $2.00 a month, and also teach piano lessons. The Turkey Match was not successful as a snowstorm occurred all day. The Morgan Post Office and Telegraph Station have been moved to Eagle Mills.

It is now December 20 and nothing about Christmas ads or sales in the paper, except an ad for the A. A. Anderson Jewelry store in Ishpeming. Alex Heyn is back in business and the paper notes that he has filled his store with toys. The Methodist Episcopal Church is having a difficult financial time. It built its new parsonage in 1872 just before the "Panic of 1873." and is hopelessly involved in debt. All those owed money are asked to consider what they would accept. Children at several Sunday Schools who were serious attenders were given very nice Christmas gifts. The year ended with mild winter weather again. The buds on the trees are getting quite large and woodcutters are saying that sap is running freely. Doctors report two or three serious cases again of scarlet fever. The Teal Lake ice which skaters were enjoying has entirely disappeared. The meat at the meat markets is being lost, and five carloads of meat going to the Copper Country were unusable.

1878

The year starts off very warm yet. On the corner of Jackson and Cyr Streets, the editor notes that pansies are flowering. Pastor Wealen of the M. E. Church announces that all of the debt has been paid in full. It also starts off very sadly with the death of one the infant twins of Dr. and Mrs. Cyr. Another horse disease has entered into our area that causes horses to become paralyzed before dying.

17. The Terrible Explosion

The news that will be remembered and written and talked about the most will be the large Nitroglycerin Explosion. Nitroglycerin was being loaded into a C&NW railroad car by the north Jackson Mine Pit, where the tracks leave to go to the Teal Lake Mines of the Bessemer and Cambria. 4,800 pounds of the Lake Shore Nitro-Glycerin Co. explosives were being packed to send to the Republic Mine. Seven men were killed. Of three, the only body parts found were a couple of pounds in weight. The train engineer, the fireman, and two brakemen were found at the bottom of the locomotive cab in a very mangled condition and burned and unrecognizable. One body had only a partial skull left. The large steam locomotive and its tender were sent about 50 feet down the track. Glass windows broke at the Jackson School and in most all homes and businesses of the area. Ceilings fell and shelves came down in stores. Furniture buried people in their homes.

Although Negaunee Miners at work were violently shaken underground, there were no cave-ins. Captain Merry had walked by the area as the nitroglycerin was being loaded and he thought that they were doing it rather carelessly from the sounds he heard inside the railroad car. He was behind a hill when the explosion occurred but was still thrown to the ground by the concussion. At the scene, 50 feet of track rail was removed from ties and lay twisted in the area with other debris on the snow. There was no sign of any ties at all as they were reduced to slivers. The train wheels and tubes and bell were all in scattered parts. The hole was five feet deep and 25 feet in size. No one is living to tell what happened. People in Chocolay Township felt the ground shake, and items moved inside homes five miles away. The date of the explosion was January 2, 1878.

There is a heavier man around town who forces his small dog to pull him on a sleigh, and uses his cane for a whip. Of the two, the dog is more civilized. There will be a race for all dogs pulling sleighs, with a new sleigh for the winner. Charlie Griffey and Willey Webb won the races. Lots of dogs got tangled up in their harnesses. There is more smallpox news. Two people from Canada have brought it to the Tenant House at Case and Pioneer. There are now ten cases by the end of January, all in the Tenement house area. Charles Demerce, 14, has died. Because of the smallpox outbreak schools have been closed. Names are printed in the paper. All are under 23. In addition in January, several cases of scarlet fever are reported.

In February, the snow has about disappeared again. The paper notes that the snowplows on the trains are for decoration only. No ice is able to be cut on Teal Lake, and ice skaters are warned to stay off. Cows are out to pasture. Summer clothes have appeared quite early. School has restarted and all children are required to get vaccinations. Six of twenty smallpox patients have died.

The newspaper notes that it might not be good to get ice from Teal Lake as all the slaughter house remains are put in it, as well as all the refuse from the nitroglycerin plant. It was thought the fishing might be affected, but there are more fish than ever this year. Gold is selling for $102.50. Mr. Philip Feibisch, who left here three years ago, has returned and started up his business again. John Harris left here also and is now farming in Barton County, Kansas. The fire department heater for the Steamer truck is working well, but they need a new firehouse with a tower to dry hoses out. They have lost 500 of their 1500 feet of hose from rotting. More people are learning to read English and more are reading and so Kirkwoods have opened up a bookstore. John Sawbridge has opened up the third hardware store in town. The Lonstorf Glycerin Co. in Humboldt has sold out to the Lake Shore plant here and all will be moved to Humboldt. Negaunee is happy. Something should be done to stop the manufacture of poisonous wallpaper, made with colorful inks containing arsenic. Good Friday was generally observed as a holiday with all of the mines closing. Seven children were born on Easter day.

Tim Curran and partner John Arnott had two falls of rock fall on them at #7 Pit. Tim Curran is killed and John Arnott hurt, but improving. A circus came in May, and attendance was pretty good for cold and rainy weather. The editor noted that "the performers were visibly cold and awkward in their routines." Kirkwood's store has put in an elegant soda fountain. The Clifford's players came twice to do plays for the appreciative Negaunee audiences. Rev. Miller of the Episcopal church resigned because of the poor economy and church finances, but members got together to raise funds for his salary for another year. H. J. Colwell left here and now has a cheese factory on his dairy farm in Kansas. A lady in Escanaba was found dead of "a visit from God hastened by the use of alcohol." The Catholic Society is cleaning up the stumps and brush in their part of the cemetery. Boys continue to stand on the streets and insult the ladies. Mr. Breitung's son, Eddie, has the only pony in town, and a cart to go with it. Wild pigeons are numerous and people are enjoying some nice pigeon meat pies. A family, recently from Denmark, has brought smallpox here again.

Negaunee beat Ishpeming at baseball again and won $20.00. People are wondering if there is going to be a revival of the iron ore industry as the Jones and Laughlin Steel Co. will not close its mines here for the traditional two-week vacation this year. The largest circus currently in the USA, the Forepaugh's Great World Circus, will be here in July. They will come in 34 double-length Palace cars, and will have a Circus, Museum, Aquarium, and Menagerie. The show arrived and a total of 18,000 people came for the two shows. Many were turned away. All was as advertised. It was accompanied by a "hoard of thieves."

If the amount of dog taxes totaled the total of dogs, we could build a new firehall. A hard-working miner has a wife with small children who was seen reeling around town for several hours, and spectators finally picked her up and carried her home. The Iron Bank of Capt. Mitchell has gone into suspension until he returns to town. It is thought that depositors will receive their money. Mitchell did return and on September 12 paid all depositors dollar for dollar and then the bank will close. It is a great loss to Captain Mitchell. One fellow who did a lot of complaining for a month actually was found to owe two cents.

18. The Miner's Powder Co. Great Explosion

The Powder Company that moved to Negaunee not so long ago has also had a great summer explosion. Four men died. It was located just 1.5 miles southwest of Negaunee and was located at Mud Lake's eastern shore (near Cliff's Drive) and thought to be caused by Mr. Hubor processing a new type of powder he invented that does not use nitroglycerin and could be smaller in size. He died along with a Mr. Scanlon, Cooper and Brown. The plant was quickly rebuilt. When completed, the company will only make Vigorite, and not the material involved in the explosion. A nice article is in the August 22, 1878, Iron Herald.

19. Cholera and Diphtheria

Cholera has lately taken the lives of several children under age one. Everyone seems to be working. There are no extra rooms or homes currently, and no idlers around. Mr. Atwater's Hardware store has gone bankrupt and the Sawbridge Brothers have purchased his stock and building. What has happened to the Poultry Association? Mr. Foley has lost a team of oxen, a young heifer, and now a fourth animal. He believes someone is helping them to get lost.

Mr. Carr lives in the back of Charles Sundberg's Jewelry Store and heard a burglar. The burglar shot at him and grazed his side, and then escaped. The constable found a man had just bought a pistol from Mr. Neely's store that morning and promptly arrested the burglar. Wm. McComber of the mine by the same name has disappeared, and was later found to be in Chicago and Cleveland, trying to get a partner for his mine and pay off debts.

Two more children have died of diphtheria. There can't be many families not acquainted with death in Negaunee. There are seven students at the Ann Arbor University this fall semester. In October, four diphtheria deaths, and two more have it. People are asking for places to rent. No signs up anymore. Prof. Fohrman obtained a piano and his pupils gave a concert. The town gave its thanks by not showing up. He should have advertised it as a Negro Minstrel Show. The Bell Telephone Co. is putting in telephones for the Iron Cliffs Co. between the Salisbury and Barnum Mines and the Pioneer Furnace. The Cliff's Furnace on the Cliff's Drive has closed and a large blower will be moved to the Pioneer Furnace stack #2. It is now December and the paper notes that those lads who bade good-by to Sunday school right after the fine summer picnic, will no doubt now return for the Christmas program. Wm. Brand was killed while walking on the C&NW tracks. The train wheels "passed over his head, crushing and mangling it in a sickening manner, his face being distorted beyond recognition."

Mrs. L. Ryan sold her cow by lottery. 303 people bought tickets and enjoyed a fine ball after the affair. Helping Mrs. Ryan, noted the paper, was a good charity for a fine person. Ore production, in 1878, in order of the largest to smaller mines was: Republic, Cleveland, Lake Superior, Jackson (83,000 tons), Champion, Michigamme, and Salisbury. It was nice shopping this Christmas. Wettstein's Jewelry Store bought a thousand dollars of jewelry from a failed store and has offered good discounts. Watches that were $50 to $100 are now available for $15 to $20. And the last news of the year was that the Marquette and Mackinaw Railroad will have its property revert to the state as a no-start company.

1879

A. J. Mead has closed his Negaunee store and brought all of his goods to his store in Humboldt where he will live permanently. The Rolling Mill is now filling with water as the pumps have been removed. The Palmer Mine will be sold at a Sheriff's Sale. The new outdoor ice rink is finished and the cost for the winter is $4.00 for men, and $3. for women. Mary Foley was unconscious for a bit last week when she fell on her face at the end of "crack the whip." The Ishpeming Cornet Band was on our streets on New Year's Day playing wonderful music. The Peninsula Brewery on Gold Street will be reopened. Little Henry Bennett, 9, has died from diabetes. Lars Peterson of the Jackson Location lost a child when a pot of hot coffee fell on him on Christmas Day. The library issued 4,659 books last year to patrons.

The state legislature has given the M&M railroad more time to build its grade. The youngest of the Tom McNabb children died of whooping cough on the 18th of February. Europe is having trouble with "The much dreaded Plague" as it moves westward from the shores of the Caspian Sea where it first appeared. Frank Boone, in prison for robbing Cyr's Drug Store, may be paroled because of a citizen's petition asking that he come home to aid his mother with support upon the death of his father. More than 70 couples attended a masquerade ball at the Excelsior Club. 60 people have passed through Negaunee on their way to Leadville, Colorado, and 19 more from the Copper Country are to follow in a few days.

Another death in February from whooping cough. William Nast has returned to town and will take over the business of H. E. Mann. The price of iron ore has gone up and the Jackson Co. has given all its employees a raise in pay. A steam wagon made its appearance here in the county. It travels on normal roads and can pull large loads. The St. Patrick Society dressing in their finest green held a parade which was observed by the general public as well and then a church service was held with Rev. Father Vertin officiating. Dr. MacKenzie is concerned of several deaths in March from diphtheria, where funerals were held involving two families at a church with the general public attending. No doctor was involved and no one had knowledge of the illnesses at the time of the funeral. His concern for the spread has led him to ask for signs to be posted in both English and French to not enter homes where the disease exists.

Fitch Bros. Groceries is closing and the men are leaving for Leadville, Colorado. We wish them every success. A former resident has written from Leadville saying that they now have 30 hotels, and 15,000 residents. Included in the businesses are many dance halls, gambling houses, and concert theaters. A tunnel from Canada to Detroit will be built under the Detroit River. Ed Breitung has finished his term as state senator and has now been elected the city mayor. James F. Foley has had 5 lads arrested for using his barn on Teal Lake Avenue as a gymnasium. One of our well-known young men, Patrick Donohue, is leaving to visit his brother in Iowa, and then going on to Deadwood in the Black Hills area. The Rolling Mill property has been leased and pumps are being installed again to reopen the mine. Tom Tracy has a crew mining at the Allen Mine property. A. C. Seass, our fine, upright, cigar maker and tobacco dealer has sold his business to the Feibish Co. and he is leaving for Colorado. The MH&O and the C&NW are busy surveying to bring their tracks to several new mines south of the city.

20. The "White House Burns"

At noon Monday morning, bells and whistles sounded and the sky was lit up by the Bessemer Mine, which was thought to be on fire. It was, however, the "White House," a large building on the shore of Teal Lake, and now owned by the Teal Lake Mining Co. No help was available to put the fire out as it was too far for the steamer to come. The rich man and his daughter built before the city was named, and had "rich upholstered furniture and magnificent carpets." All was left there to be carried off after the time came when they did not return. The building even became a barracks for troops at one time. It became a place for fishermen and hunters, and orgies as well, stated the May 1, 1879, issue of the Iron Herald.

The Cambria Mine is being pumped out to begin mining operations. The Sawbridge Bros. have erected a large Tea Kettle over their store front with their name on it. There was a snowstorm on the fifth of May. The Vigorite Powder Co. has sent several orders to Leadville, Colorado. Dr. Cyr has also installed a fine soda fountain and "will put it in blast on Saturday." A young Negaunee man, enrolled at the University at Ann Arbor, has been arrested twice, the first for burglary, and the second for escaping from jail. The Manganese Mine has received its railroad grade and will begin shipping soon. Walking contests have become the rage, starting with short distances to ones of several hours. Someone suggested putting cinders on Iron Street. Bad idea~~ it needs a hard coat of gravel. The Edison Phonograph Company is in the area for a week seeking people to talk into the machine and leave their children a record of their voices. Few customers and no idea yet that it can be used for music. The walking contest is around the edge of the ice rink. Seems like an odd place to be walking. It is 231 feet around each lap. Tom Flynn won, taking the lead at lap 80, and finishing five miles in 38 minutes.

A large number of immigrants have started to arrive here in the area. 60 people came in the last week. Deer hunting will be from October 1 to December 1. The Ishpeming Barnum House burned down and several other buildings with it. Mr. Seass only got to Kansas where he bought 640 acres in Ellis County. Three miles south of Cascade, Sweitzer's Mill burned to the ground on its tributary of the Escanaba River. Four of six houses at the Baldwin Kilns area have also burned down. It's June 5, and a heavy frost came the previous Monday night. The next Friday morning, there was considerable ice on sidewalks. Then, warmer weather brought many people to bathe in Teal Lake. The second walking contest was a failure with no contestants and few attenders. However, some boys staged a contest with the winner getting a dollar, and they all had a fine evening.

A long article of how Negaunee needs a good hotel. There is little hope for the Morgan Furnace to resume as the trestle tram road burned to the ground. On Teal Lake, a speckled trout weighing 3 ¾ lbs., was taken in the docks area near the boat houses. D. E. Patterson is now in Colorado with his "Cardiff Giant." H. C. Coburn owns the Giant and reports good exhibition business. Republic has its new Town Hall. A Case St. gardener had Mr. Griffey come and see the frost damage to his peas. Mr. Griffey didn't think frost hurt peas, and the gardener "twisted the corners of his mouth" and said, "They were bitten by a cow." The city is celebrating the 4th on the 4th and 5th this year with a parade and sports contests of horse racing, baseball, and "pedestrianism." When the day came it was terrible weather and the Driving Park was covered with water and mud. All was held, however, as a patriotic celebration. P. B. Kirkwood purchased two beautiful male parrots, but this week he found eggs in the cage. Richard Eddy, 40, died at the #6 Pit when a wall of soapstone detached and buried him. He leaves a wife and five children. Mining operations were suspended at the Jackson to remember Eddy's death.

21. The Breitung House

Mr. Seass sold his new business in Denver and has returned to Negaunee. He has secured the Ogden property for the new planned hotel. (August 14) The new hotel at the head of Iron Street will be three stories. The entrance will face Pioneer Avenue, with a ladies' entrance from Lincoln. Lots of new buildings are also being built near Republic at "Pork City." On August 14th, a very heavy frost killed all tender vegetation, including some potato tops. Hematite ore from the shafts at the #7 pit contain 67% metallic iron. The City Common Council has prohibited auctions and hawking on street corners. The Marshal Jeffery has collected the poll tax from 506 people. Now they can vote. Rev. Vertin of St. Paul's here has now been consecrated as the Marquette Diocese Bishop. The Carney Stable on Iron Street will be auctioned off. Mr. Carney is currently in the Wayne County poorhouse, after suffering various reverses and misfortunes. It is now the end of September and Mrs. Seass anticipates that the new hotel will be ready for guests in early December. The building will be enclosed next week. R. M. Werner has returned from the Black Hills and will re-establish his business of making and selling candy. Tom Ryan bid $550. and won the Carney Livery Stable. The Methodist Episcopal Church has 1,570 full members in the Upper Peninsula in 21 churches.

22. Electric Lights

A new first is that the Ishpeming Lake Superior Mine, in 1880, has put electric lights in three pits. Some boys, age 8 to 15, have been stealing from stores. Three had their pockets full as they left Dr. Cyr's establishment.

23. Another Fire in Town

An "extensive conflagration" visited the city last Saturday. (October 9th paper) The fire embraced the south half of the buildings between Gold and Silver Streets between the alley on the north and the MH&O tracks to the south. It began in the railroad depot where only records were saved. The south wind made it a ball of fire and then lit the neighboring buildings up as flames went north. The fire was put under control after the Jackson Mine fire department arrived with the hand pumper to help out. McKenna's Livery Stable, the Central Hotel, and the C&NW depot burned. Three Fuch buildings, Junction House, Wheelock and Suess, Gaffney Building and Furniture, Kuhlman and Brown, Mr. Masse Cigar Co., and several other stores suffered great loss. The Ishpeming Department arrived at the fire's height and the city is extremely happy for its help. The entire block of buildings were ashes in about an hour after the alarm sounded, but the rest of the town was saved, including several homes "a quarter of a mile away" that caught fire from hot ashes.

An eleven year old girl has died from scarlet fever. Dennis Smith, riding on a locomotive in the yard area, fell off and lost his leg to amputation. Harry Fuller won the cow of the Stuart's raffle after buying only one ticket. 250 Frenchmen from Canada have arrived here and settled in. They will all work in the woods this winter. There will be a new bank here called The Michigan Iron Bank. The demand for ore is so great that one of the stockpiles at the Green

Bay property, mined 7 years ago, has been shipped. In early November the Hiberian Blondes appeared at Sterling's Hall and people attended to see the "gyrations of the nymphs." It was just a "saloon variety show." Dr. Cyr has put a street lamp outside his store. (gas?) The paper suggests others might do the same. A new church has organized as the "Presbyterian Society," and is proposing to build a church. Mr. McKenna has auctioned off all of his livery equipment, including horses. He will re-supply when his new livery is completed. A new mine is underway by Adams and Foley. It will be called the Milwaukee, and a road will be put into the property in the spring.

There was a bad snowstorm at the beginning of December and the Pioneer Furnace whistle sounded and continued to shriek so that residents thought the worst was happening. The storm made the valve stay open. Work stopped for two hours as the steam had been used up and it worried the company that the boilers might blow up. The weather for the Turkey Shoot was okay, but the shooters were poor and only a few birds went home with winners. The Breitung Hotel is grateful to the C&NW for the free building materials transportation. For the first time, the pews of the M. E. Church have all been rented. The new fire engine house is completed and the steamer has a new home. The presence of Susan B. Anthony at the C&NW depot Saturday afternoon drew quite a crowd. She had lectured in Marquette.

—A Kansas teacher has set the following afloat: A cowboy has three ponies and a Mexican saddle which are worth $220. Placing the saddle on the first pony makes it worth the other two; placing the saddle on the second pony makes it worth twice the other two, and placing the saddle on the third pony makes it worth three times the others. What is the value of each pony?

1880

Negaunee proudly held its annual dog race, but was sorry to see Ishpeming carry off the prizes. No one died in the great town fire a few weeks ago, but Mrs. Cohn lost everything except the clothes she was wearing. Through kindness she has found a new home and will sponsor a grand ball at Winter's Hall. The Amateur Dramatic club will give a play. "Girls, it's leap year again, so go hunting, and get your license when you're done." Three mines have now offered Benefit Insurance to their employees who will pay 40 cents a month and get $30 a month if disabled, and $400.00 in case of death.

The Detroit, Mackinaw and Marquette RR is finally getting underway when equipment arrived in town for building it. There is a shortage of horses and not enough can be brought here and sold. Some cost $800. The Cardiff Giant has been sold by Mr. Coburn for a mining claim in Colorado. The Humboldt Iron Co. has increased the wages of employees to $1.75 a day. The C&NW is now using only coal-burning locomotives. John Best's cat fell into his well. In retrieving it, the other items found were: a rat trap, a pair of cow horns, a breakfast plate, an assortment of stockings and mitts, a number of oyster cans, two pails and more yet. He also took out 40 pails of mud. Tom Taylor shot a large white owl from the chimney of a Cyr Street residence and gave it to a taxidermist. It has a 5'4" wingspan.

We find that the investors of the new hotel are Ed Breitung, Henry Merry, and Joe Kirkpatrick, and that they have run out of money and the hotel is not yet done. They are now selling $8,000. worth of bonds which will pay for furnishings as well. A white owl has also been held captive in Dr. Cyr's cellar where it has caught many mice and rats. It's a real mouser!

Miss Jennie Jones, upon leaving a church service, was run over by a team of horses, and survived with only a lacerated wrist. In lieu of St. Patrick's Day activities, the money will go to help the starvation in Ireland. Dr. Cyr has a one-horse snowplow which works wonderfully and the city should consider getting one. The hotel bonds will bear seven percent interest. 50 Scandinavians have come through with most going to Republic and mines west. The East Iron Street bowling alley has been converted to a mine ladder-making company. Thomas Flynn, an Iron Herald employee, has been named the city librarian. Tim Donohue plans to move back here from Iowa as soon as he sells everything. There is no room in town for another person. "The town is absolutely running over." It is now April. A live goose was found 400 feet down in the bottom of the Pioneer Mine shaft. It had to be there for ten days without any food. . The "eating house" at the Union Depot has been opened to the public.

24. First Surface Cave-in

Over 20,000 tons of ore inside a bluff between #4 Jackson Pit and the Pioneer Shaft House caved into an area about 30 x 70 feet. The bluff had been undermined and supported by an ore pillar which crushed. People who saw it happen said it was a "grand fall." Work had stopped in the area two days before and the cave-in anticipated. The ore will be mined from surface rather than from the underground. The following week a torrential rain caved in a shaft at the Jackson just west of the # 7 pit. Peter Mantague, recently from Canada, fell down a shaft at the Pioneer Mine just before his candle was lit on the first level. He fell 110' to the bottom and into some water which broke some bones, but saved his life. Several men working on the railroads have had serious accidents, but no deaths. In May, two railroad men died, one by tripping on a rail in front of a train, and one

falling off between cars. The C&NW has delivered 236 railroad car-loads of freight to Negaunee in the past month.

The Tom McKenna family lost an infant to whooping cough. It is the fourth child they have lost. Two coaches of people direct from Sweden have arrived in the county. Wm. Jeffrey died while loading a car with a wheelbarrow at the #5 pit when the track gave way. Telephone lines are now going to Cascade and to Marquette. Colic has killed two $300 horses. Meningitis killed a six year old daughter of Phillip Marchette. The July 4th celebration this year will have a horse race open to all horses. The city treasury has about $6,000 beyond what will be needed by the end of this year. The MH&O have built a fine new passenger coach and it will not be red, but sky blue, a pleasant change. On July 4th, five horses entered the horse race, and horses from Marquette and Champion won the top two prizes. The old building next to the Iron Herald has been purchased and will be removed. It was built in 1858.

A girl with the Billy Maple Theatrical group, which played here, became ill. They have left her behind in Ishpeming and some ladies have cared for her and are arranging to help this frail girl with a spine problem. Cascade had a grand 4th celebration with Joseph Kirkpatrick giving candy to 130 children and fireworks launched from a high 15' balloon. Some lots are being sold on Peck Street as water is available only 25 feet down. At the new hotel being built, the dining room is now being completed and some dances are being proposed to benefit the project. Tobin Street is now open to the bluff on the north. To end the quarrel between Republic and Negaunee as to who has the best baseball team, a game with the winner getting $200.00 will be played. 25 to 30 Finnish people have arrived direct from Finland to settle in this vicinity. There has been a cave-in on the east end of the Pendill Mine before men entered for morning work, 50 feet in length and 12 feet deep. Negaunee won the game with Republic, 21 to 33. A chartered train carried the team and fans for the game. Paint manufacturers looking for red dust for their paint should have their attention drawn to Iron Street, which has an inexhaustible supply. Jerry Vaniderstine has returned to Negaunee from Leadville, Colorado, saying the climate did not agree with him. Two men went to Lake Michigamme to explode some powder and pick up a few fish, and Tom Rook is now missing his right hand. The last bagnio has been closed up in town. Sealskin coats with beaver collars are the new fashion and will sell for $500. The Alfred McCombers have lost a little boy, and Daniel Kelly, 14, died from the measles. Louis Heyn, a fine merchant here in the 70's, returned for a visit as a salesman for a Milwaukee clothing house. A fellow who says he is the nephew of Kit Carson will be here and giving open air lectures as Kit Carson, Jr. Professor Dare was also here doing tightrope walking and collecting nickels and dimes. He got about $25.00. Finally, the Jackson Iron Co. has purchased a diamond drill. They have been busy digging exploratory shafts up to now. Charley Mitchell has been having an ill feeling of late and he has had a 50 foot tapeworm removed. The Negaunee Fire Department is asking the Common Council to increase its membership from 30 to 50.

A small school will be built at the Rolling Mill Location, there being about 40 children there. Train 15 hit a cow early this morning and broke a leg and tore off a horn. The Iron Cliffs paid out its monthly payroll to about 1000 men and it "makes a fat streak in the business on Iron Street. At the new hotel, the work is now being done on the third floor. The Winter and Suess delivery wagon and team have had two run-offs in the past two weeks with wagons being destroyed. One horse of the team also had to be shot this week.

Dr. Cyr has a rabbit and a fawn that eat out of the families hands, and the rabbit follows the fawn all day and lies under its guardianship at night. The Pioneer Furnace hopes to keep both stacks going until shipping is over for the year. It is now October 7, and Stack 2 has run steadily for 14 months with no signs of giving out. It has produced over 12,000 tons of pig iron. The autumn Teacher's Institute has met and the teaching will compare the old style compared with teaching in the present day. Carl Gustave Sundberg has arrived from Sweden to live here near his brother, Charles Sundberg. Frank Sylvester is building a small home at the corner of Teal Lake Ave. and Clark Street. Charles Naumann is building an addition to his pop and root beer factory. A large number of Finnish people have been arriving and are finding work in the mines. A young man came to town and was later found in some bushes with a bullet in his chest. He said he was unable to work and wanted to die. He came here from Canada four months ago. Capt. Pascoe called his mine officers to a meeting at Republic and they feared the worse. Instead the company rewarded them by dividing $1000.00 of gold coins among them. The hotel, almost a year late, is now receiving some guests. A week later, it is official. The name of the new hotel will be the Breitung House!

The DM&M railroad is now complete from Marquette to Deerton. At Republic, the local girls enticed a lot of boys to come to their leap-year party. "They came by squads on every train" and girls met them at the depot and provided supper at the Ely House. The young men had even been growing mustaches for the past few weeks in preparation. Because of a shortage now of wood, more people are buying coal stoves. The #1 Stack of the Pioneer furnace has blown out after 11 months of steady use. Stack #2 continues making about 30 tons a day of pig iron. Some Belgians arrived in town with wooden shoes in November, and they could not find anyone to communicate with, and left town "in a pitiable condition." Manufacturing butter out of lard has become a regular business. It's counterfeiting. Water flooded a drift in the # 7 pit and now the wells in the west part of the city have dried up. Marquette County has 25,363 people, largest in the U.P. It's Christmas, 1880, and the C&NW carried 483 packages this week.

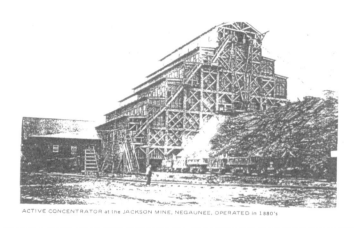

ACTIVE CONCENTRATOR at the JACKSON MINE, NEGAUNEE, OPERATED in 1880's

Here is a photo of the large Concentrating Plant that failed after many attempts to operate efficiently. It was located Southwest of what later became the Mather B Mine. Econ. Develop. Corp. of Marq. County, 1991

Here is a photo of the Cleveland Cliffs Iron Co. Bellevue Farm on the road between Negaunee and Palmer. Photo is courtesy of the Negaunee Historical Museum, from the Gazette of Sept. 20, 1972, and courtesy of Joseph H. LaPointe.

1881

The year started off quite cold as on January 10, at six a.m., the Negaunee thermometers read 38 below, but some of the places in the western county read 43 below. Many wells are becoming dry. The newspaper notes that without Partridge Creek that there would "be virtually a water famine in this city." A deposit of iron ore has been found on Section 31 of T48R25, five miles west of Marquette near Morgan. Alexander McDonald, who lost his sight in the McComber Mine accident, has gone to Chicago for help as some sight is returning.

25. The Concentrating Plant

A company from New York has decided that they will build a stamp mill to break down iron ore, and it will be built between Teal Lake and the Jackson Mine. It will use lean ore previously put on rock piles and discarded. Now the iron ore will be saved. Mr. Conklin represents them.

The Presbyterians are hoping to build a new church on West Iron Street. In February there was a very violent snowstorm, stopping all trains and lasting 24 hours with many snowdrifts over ten feet. Wm. Hodge lost a leg at the Pendill Mine. A fine sketch of the city of Negaunee has been made by H. W. Fletcher. Each is $2.50 and one can see almost any house. At the Breitung house, elegant banisters have been added to the stairways. The newspaper requests that the many smokers at the post office think of the ladies and non-smokers and do their smoking elsewhere. The Chicago Mine is being readied for a full operation this summer. A suit has commenced regarding the 12 shares of stock given to Margi Gesick for revealing the iron ore deposit of the current Jackson Mine. Another large snowstorm came at the end of March. It took four days for the Chicago and Northwestern train to get through on a regular run from Chicago. It had four days of mail aboard. There was too much snow for the annual dog race.

The March 31, 1881, newspaper has the amended area set aside for the city of Negaunee with a legal description and "declared a city by the name of the City of Negaunee, by which name it shall hereafter be known." The town now has 3,931 people. Ishpeming has 6,039, and Marquette, 4,689. More snow on April 13. Two families have a tunnel to their outhouses. The Iron Cliffs Co. hired a gang of men who worked all day clearing Iron St. in order to keep ore moving from the Jackson to the Pioneer Furnace. Mr. Johnson has a new building at the corner of the MH&O and Gold Street. The library is requesting the moving of the library from the town hall to the upstairs of the engine house (fire hall). Ben Neely will build a new building at Gold and Iron Streets. The celebrated singer, "Litta" gave a fine concert in Marquette. Ore has been discovered near the Baldwin Kilns. Peter Trudell has rented a corner of the post office where he will sell stationery, confectionery, etc. For 20 cents, one can get a list of all the library books. The Breitung House has been averaging about 60 or 70 guests a night.

Mr. Conklin has gotten contracts with several of the local mining companies for "lean ore" to be used in their concentrating plant to be built. Mr. Henry Lee has raised enough money from businessmen to use a sprinkler to put water down on Iron Street. It looks like the Pioneer Mine has gotten out all its iron ore and will be closed by the Iron Cliffs Co. L. Marcoggi fell from his wagon while turning a corner and is paralyzed from his waist down and has been put in the county house. The only tent show to come this year is Beckett's Circus and it is here today. The complete end of the Morgan Furnace occurred, when all that was left and burnable,

caught fire this past week. All the machinery had been removed some time ago.

Two men preparing to bathe in Teal Lake found a dead infant. The doctor says it had been born alive. Harry Rowell was arrested for an assault on a ten-year-old girl. Two boys, in separate incidents, were miraculously saved when falling between cars of moving trains. The July 4th celebration has horse racing, pony racing, baseball, and Cornish wrestling. Money prizes for each event. The Breitung Hotel has its new porch. Four boys were arrested and three fined for throwing stones at Tom Ryan. The Peninsula Brewery is under new ownership and will have its first deliveries this week. It is located on Gold Street in back of the Pendill Mine. Mrs. B. McCarty had triplets. Pres. James Garfield has been shot and seems to be out of danger. Assassin arrested.

26. The Johnson Sawmill

Mr. Isaac Johnson will move his sawmill from Little Lake to Negaunee and will be at Teal Lake by McComber's dock. He will make sash, doors, molding, blinds, etc. If you don't want your family to get diphtheria, roast your own coffee and the pungent aroma will destroy the germs of the disease. The Wizard Oil Show came to town and had great free music and entertainment. Notes the paper, "They also sold a lot of oil." Mr. Marceau died in Clarksburg. He had been married 3 times and leaves 27 children.

It is now August and the Lake Superior Carriage Works on Silver Street has a booming business. They plan to build 75 sleighs for the winter. The Fire Company No. 1 was out practicing hooking and unhooking hoses. Mr. Johnson has his saw mill machinery here, and is building a large boarding house, and a carpenter and blacksmith shop. He has 600,000 feet of logs in the lake ready for cutting. President Garfield's life now hangs by a tender thread. The roof of the tunnel is being removed at the Jackson #7 pit.

27. A Water Works

On Sept. 1, the newspaper makes the first suggestion, in lieu of the water shortage, that a Water Works be built on Teal Lake to supply the homes of the city. There was a bad train accident, but no deaths, of two trains in front of the Union Depot. The wife of Prof. Zara, a ventriloquist, died here after being ill for three weeks. He went on without her to keep his show schedule, but has been back to see her. She evidently is buried here. James Foley is building a nice residence at Case and Teal Lake Ave. by the Methodist Episcopal Church. Someone broke the boom of the logs at Mr. Johnson's sawmill, and logs are now floating all over the lake. Charles Thoren's youngest son, 2, has died. Pres. Garfield has died. A heavy rain estimated at six inches fell. More rain in October caused mines to be filled with water and mud of some quantities that the pumps could not handle. Rowdies took store signs and threw them in the Jackson #7 pit, and also opened up many gates, allowing cows to enter and eat several gardens. The Catholic Fair had a profit of $871.60. Tom Mylivoja, who came from Finland just two months ago and was saving money to bring over this family, died of typhoid. The DM&M RR has been completed to St. Ignace. "Porkopolis" (Pork City, west Republic) is now dignified by the title of Park City. (November 24.) The city will pay physicians to vaccinate all people for smallpox who need the bovine virus. 50 carpenters will soon be hired to build the "reducing plant." It is expected that 500 or more men will be hired when completed.

Christmas at stores was quite good and some Iron Street dealers made $1000.00 or more last Saturday.

1882

The Reducing Plant is having a 500 hp. engine being built and the building will be the largest in the Upper Peninsula. Lumber for it will cost $50,000. 40 carpenters have arrived from N. Y. Total workers are now about 100 and have received their first payday. Railroad tracks are being put in. The Jackson Mine will soon end its hauling of ore to the Pioneer Furnace by teams up Iron St. and begin using the railroad. Joe Lampshire died from a premature blast at the Lucy pit of the Bessemer Mine. The Milwaukee Mine had a total fire and Mr. Foley has now ordered all new equipment. A Canadian couple with two children moved here a few months ago and have been living in squalid conditions and the wife has died of typhoid. Some kindhearted women have prepared the body for burial. The Catholic community has purchased land at Pioneer and Case for a school and home for the Sisters. Phil Kirkwood caught some trout through a hole on Teal Lake. Doctors have vaccinated 405 public school children in town.

With many new mines opening in town, and with the new concentrating works, there will be a real shortage of housing here. Quite a few men were injured in mining accidents and street brawls in April. The Victoria Loftus British Blondes will perform, "Little Sins and Pretty Sinners" at Winter's Hall. The show was actually pretty good. A new home will be built at the corner of Main and Teal Lake Ave. by Mr. Joseph Winter. More details of the Concentrating Plant: It will have 225 men building it, at 183 feet long, 116 feet wide, and seven stories high. Tom Pigott has started milk wagon deliveries and some others as well. The Iron Cliffs lease of the Pioneer Furnace and the Jackson Mine will soon end, and in the future the furnace will get its ore from the Salisbury Mine. A brickyard on Section 8 in the south part of the city has reopened. The Concentrating Plant is in need of more water and requests the use of Mud Lake and the Partridge Brook.

The Concentrating Plant paid $20,000 in wages and Iron Street was booming until almost midnight. The Plant also is requesting the city for a road to their place. A ten-ton crusher fell 4 stories in the building when a hook broke. 30 dwellings are currently being constructed. There are now 1407 children of school age. Mr. Neely's new building will have an opera house. A stage, largest north of Milwaukee, a circle gallery, and box seats will be included. In June the idea of a water works is getting positive attention. The Concentrating works will also have a foundry to make iron castings for local use. The Breitung wants a water line from Pit #7, but the city would like it to serve other places as well. Parts of the Fire Steamer have been sent to N.Y. for repairs. It was damaged from sabotage. Mr. John Carkeek won the wrestling tournament on July 4th. The concentrating mill is about to begin crushing ore and has spent $250,000 this far. The water works project has gotten a black eye from the city charter which does not allow any more than $29,000 to be spent of taxpayer money. Two men were put in jail after arriving from Marquette using a horse with blood and sweat pouring out of him and scarcely able to move. The C&NW killed seven cows on Sunday evening, four inside the city limits. A road will be built from the Champion Mine to the Republic Water Works. Moses Goodrich was badly burned at the Pioneer Furnace as he was about to pour ore in at the top, and a belch of burning gases came out and caught him on fire. The Presbyterians have begun building their new church at Pioneer and Case Streets. An August frost killed some tomatoes and cucumbers, but the gardens in town escaped. The C&NW is going to add onto its tracks from Ishpeming to Michigamme. Mr. Antoine Barabe has bought a new mowing machine and many go out to watch it. Abbie Fitch left here three years ago and has traveled all over the west. He has returned and plans to stay here. The people of Humboldt will build a church. The Iron Herald has received a power press to print the paper. There is a new ordinance against geese because several families have some wandering the streets.

28. The Roman Catholic School Opens

In the September 7, 1882, paper, it is noted that the public schools started with a more slim attendance this year. The new Catholic Convent School opened the same day, "at which there was an attendance of 316." Pastor Eis had a "thank you" card in the paper thanking people for the donations to erect the St. Paul School building.

The Horse Racing events are getting larger. The last races were won by horses from Detroit and Marinette, WI. Four lines of telephone wires now exist between Negaunee and Marquette, with 50 subscribers. The Concentrating Company is getting iron ore from mining the tunnel of the Jackson #2 pit. The city has decided for now to lay a pipe from the water discharge from Pit #7 to Teal Lake Ave, Main and Iron Streets. They will also install hydrants for the fire department and the pipe will be a 7 inch hollow wood. A large tub has been built at the fire hall and it is the first indoor public bath room. "It should be well patronized." (9/28/1882). A caving occurred on Iron Street where the railroads cross at the #7 pit. The caving appeared underground first and then appeared on the surface, gobbling up earth and wood. A pump installed at the #7 discharge pipe will push water hard enough to reach to top of the Breitung House. Captain Merry's new home is completed and one of the most ornamental in the city. Mr. Barnum completed his mine work here and returned east where he is the chairman of the National Democratic Committee. The Concentrating Co. is having a large locomotive being built. Mr. Breitung of Negaunee ran against Peter White of Marquette for the area Congressional seat, and Mr. Breitung won by 9,000 votes in the November election. The Red Front Store lost a bushel of fine apples in a window display, when a cow broke the window and ate them during the night. Two baskets of grapes also disappeared.

The MH&O will begin building its grade from L'Anse up to Houghton. As the year comes to an end, there is thinking that the next year may not be so good. Several Rolling Mills in the east U.S. have closed down. Also disheartening are several cases of scarlet fever and diphtheria. The surface of the Republic Mine is now lit with electric lights.

Charles Thoren won a fine rifle in a Christmas-time raffle. Thoren is now looking for a dog. The Concentrating Mill has closed down for the winter. Will it open in the spring? Stay tuned. It is not yet fully built, but the investors are worried about the price of iron ore. A thought of building a bridge across the straits is being made again, and is not so absurd. Engineering is making great strides. Christmas was fairly orderly this year with less disorderly conduct, and this, in spite of the mines and railroads paying out large expenditures to their employees.

This is the Johnson (Nealy) Mill, Teal Lake, 1905. Photo from the Negaunee Historical Museum.

1883

The city has decided to have an election to amend the charter so a water works can be built. A new additional reason is that if the Concentrating Works opens in March, as planned, the water in Partridge Creek will be unusable by the public. Ishpeming has 41 saloons, Marquette 36, and Negaunee 34. Negaunee is canvassing citizens to see if it is feasible to get an Edison light plant. It might be cheaper than gas. On sections 14 and 15, north of Teal Lake, 400 feet of gold bearing rock has been found and a shaft is being put down. The foundry of the Concentrating plant has now been competed while the mill is down. Noted singer, Camilla Urso, with musicians, will appear here at Winter's Hall. Innumerable hungry wolves are around due to extremely cold January weather. There are numerous cases of frostbitten feet, fingers, and noses. Mrs. B. Carr has lost her 12 year old daughter to diphtheria. It is the third family death in the last few months. The family of Marji Gesick has been awarded about $60,000 by the Jackson Mine. It may be appealed. 49 people in town now have telephones. Simon Harry has obtained a patent for a railroad car coupling device. Because of heavy snow, a roof on a Concentrating Mill shed fell in with considerable destruction. The Presbyterian Church is completed and pews will be sold. Three fine ladies in town were found to be stealing items from the Bankrupt Kraemer Dry Goods store and had $500.00 worth out in the alley. They were arrested and then let off without persecution because of their great misery.

A post office has been established at Turin, also known as Berringer's Mill and McFarland's Hill. In March the shipping was stopped on Lake Michigan for the first time in history because of ice. Someone stole all of Mrs. Gibbs washday clothes left hanging on her clothesline. Money, chickens and geese also are missing at other homes.

Captain John Mitchell, on a trip out west with Ed Breitung and others, is having a hard time breathing at the high town of Las Vegas. The group will leave. They also expected sunshine and are finding cold temperatures. In the March 23rd paper, the announcement was made that Capt. John Mitchell had died in Los Angeles. He, and others, had developed the Saginaw Mine which was sold for $350,000. A well-known "character" has been arrested in possession of items from several citizens' clothes lines. The Breitung House is getting a veranda on the third story. It has also been renovated throughout. It has 50 regular boarders. In the last election, Mr. Merry lost the election of mayor and his backers are trying to find out why. "It's easy," said the paper, "He didn't get enough votes."

People will vote on bonds to build a water works. Insurance rates will go down with a water supply. The May 10 newspaper announced that the vote was 4 to 1 for the bonding. Also going down is the postage stamp, if you can believe it. It formerly cost three cents, but it has been reduced to two. Republic has gotten a nice new hospital. A very large fire occurred in Michigamme. The main driving belt of the Concentrator is 262 feet in length, 30 inches wide, and travels a mile a minute when running. In May another addition was added to the Mill to handle the crushed rock tailings. Machinery is also having a difficult time with the hardness of the ore rock. The fine singer, Litta, became very ill while here in Negaunee. All of her concerts locally were cancelled, and she is confined to a room at the Breitung House. The rest of the group has moved on without her. Her sister and a maid are looking after her. On May 18th, a foot of

snow fell in a strong gale. Mr. Kirkwood shot a real beauty of a blue heron at Teal Lake and has had it stuffed and put in his drug store window. The people of Negaunee held a concert for the benefit of Mme. Marie Litta, still bedfast.

The Water Works building will be built of bricks or stone at the shore of Teal Lake. Steps will be taken to keep the water pure, and the present pipes from the #7 pit will be utilized. 3,700 feet of pipe will be installed. After four weeks in bed here, Litta was taken on a cot to the train and has gone home to Bloomington, Illinois. Some concert members, and her sister, accompanied her. Singer, Miss Maggie Merry of Negaunee, has been hired by the concert company to travel with the group.

29. The MacDonald Opera House

Mr. D. MacDonald will build a new building on Jackson Street. The second floor will become the new town opera house. The Fire Co. #1 will plan the games for July 4th. They will include a fat man's race, a tug-of-war, a bicycle race, a heel and toe race, and the other regular sports. Money prizes in each category. Word has arrived that Litta has died at age 27. She was pushed into singing at a young age by her family who lived off of her $15,000 a year income from endless concerts. It is a sad story with a sad ending. Rates have been set for those who want city water. Example charges are $6.00 a year for a family of four or less. Add on $3.00 if you have a bath tub, $5.00 for a toilet, and cow, $1.00. Businesses are more. The Concentrating Co. is busy crushing rock and could do more with two more sets of crushers. A round-house and blacksmith shop will be built. Mary Mitchell, who lost her husband 12 years ago, now has also died leaving three children. The Odd Fellows are caring for the family as Mr. Mitchell was a member of the group. Republic is building a three-story hotel. On the morning of August 2nd, some snow flakes were seen in the air. Phillip Hogan has invented an automatic pop filler at his pop factory and can fill 12 bottles a minute. People are busy buying bonds to help the Water Works in amounts of $100.00 and up. There are some deaths of adults and children of scarlet fever. MacDonald's hall will be done in October and a grand ball held. The paper, on September 13, noted that a genuine, but short, snowstorm occurred. Ice formed on small pools, and all tender vegetation froze.

The Water Works pump started up and the 20 hydrants left open to clean out the piping. In October, a test was made with six hoses opened at one time and the pressure sent all of them into the air from 75 to 125 feet. Homes wishing connection will have water use this winter. The Concentrating Mill will close for the winter, several weeks earlier than stated earlier. It is believed that the price of ore is too low. They have paid out over $600,000 in wages here, always promptly. Eddie Breitung's popular pony broke a leg and had to be killed. One can now go all the way to Houghton on the MH&O RR. The paper published a news article from Bloomington, Indiana, that a grand concert was held in memory of Litta with well known singers attending to raise money for a memorial stone for her. The CCI has agreed to pay the city $1200 a year for the Pioneer Furnace use of the new water system. In November, several people left for the winter, in the USA, and one to Italy. A company wishes to purchase John James invention of a new car coupling device. Mr. MacDonald has purchased the needed stage paraphernalia required for the Opera House. The Breitung House is going to install bathing facilities with the new city water. Isaac Johnson, who owns the Teal Lake Lumber Mill, died. The funeral procession had 107 carriages in it, and the Fire Department and Ishpeming Swedish Cornet Band marched. Mr. Ed

Breitung has left for Washington D. C. to take his seat in Congress.

"Heretofore, every city and town had its own time," said the paper as it announced that all railroads are going to "Standard Time" with the country divided into 5 belts of one hour each. A traveler will be on the same time during any movement in one belt. Regarding the cleaning up of the water of Teal Lake, a road is planned on the west end so that horses will not have to cross to properties to the north on the ice, and arrangements are also being made with the slaughter house. In December, Rep. Breitung introduced a bill to make Mackinaw Island into a National Park. A study of the winter weather for the past 12 years shows that odd years bring the cold winters, and even years more mild weather, based on the average mean temperatures. The Iron Herald wishes all a Happy New Year.

1884

Two new railroads are causing problems in Negaunee. The Marquette and Western is building over many streets and damaging a lot of nice property, and the DM&M has to raise a trestle high over the C&NW in order to cross it. Dog problems also continue with one girl held by the neck by a dog until help arrived and another man was attacked by three dogs at one time. The girl survived and Lewis Corlett now carries a revolver. The nation's largest rail maker has stopped making iron rails, and now they will be all steel. Telephone poles are now going to Republic. Marquette will try a dog race. Here, in Negaunee, school teachers are being "decimated by marriages." The Pioneer Furnace made 331 tons of pig iron in one week. The M&W RR is building a trestle just west of the Merry Residence.   John Coffee, 14, is spending 15 days in the county jail for jumping on moving trains.   Several mines are reopening in June, including the Wheeling and the Milwaukee.   D. F. Wadsworth has stolen $10,000 from the Cleveland Iron Co.   The Iron Cliffs Co. has purchased a small yard engine to run through the tunnels of the Jackson Mine. Could it be "The Yankee?"

The Sell's Circus came. "Best ever," said the paper.   400 came on a special train from Marquette, and over a hundred from Republic. One accident, when Michael Curren had the claws of a black panther catch him in the eye, He got too close to the wagon. When the circus got to Manitowoc, an aerialist fell and is near death. Back in Negaunee, a lawn tennis court will be built at Main and Teal Lake Ave. Twenty barrels of beer were shipped from here to a Pennsylvania Gun Club hunting at AuTrain.   An excursion train from Escanaba brought 200 people here for a picnic at Teal Lake. Their band also came and played during the day. In August several children and infants died from cholera.   Tom Beaubier, a pitiable cripple who tries to make a living with a saloon on Gold Street, found his business robbed of all its contents. The people in Champion had a strong frost.   In September, Gold Street was opened up the full length. Teal Lake Avenue was completed to the Water Works. Negaunee has a ladies' band and they played a local dance wearing nice new uniforms.

30. The City Begins Having Gas Lights

The city has installed a nice gas light. More to follow. Mrs. Captain Merry is going out east for the winter months. Many people are still traveling by boat, and 3,598 of them went through the Sault Canal in October. In spite of the new school attendance law, many boys are loitering on the streets when they should be studying. The paper also notes that all of the fine Hungarian people who came here this summer seem to be gone. Diamond drills are being used all over town. The Tobin Street one found ore just 30 feet down. Quite a few Finns have arrived here in November and are looking for work. They worked on Canadian railroads during the summer.   When the gas company was digging on Iron Street they found an old root cellar covered up. Capt. Merry said he remembers it when Iron Street ran south of it. Thomas Knuckey raised 11 steers this summer and has sold them to local butchers. He made $500.00. It is estimated that about 300 men here are now without any work. The Jackson mine has gone two years without a death. Many citizens think the city should contract for gas lights on the main streets of town. There is now a new truant officer contacting parents with children not in school, and there has been much success.   Today is Christmas Day, but the newspaper was printed and the post office is going to be open from 11:45 to 2:15. The county is paying $8.00 bounty on each wolf killed. Children here are still dying of diphtheria. Some families currently have 4 or 5 cases of it, and it is very contagious. Some families have lost two children already.   And this sad note that two young ladies were seen drunk and wallowing about in the snow.

1885

The city had its dog race on New Year's Day this year and Wm. Eddy won. Peter Trudell might have, but the dog slowed up a great deal when the route went by his home. Other boys are entertaining themselves by getting dogs to fight each other. The MH&O RR has reduced wages. Diphtheria cases are down to 3 or 4 and schools will re-open. At 8:00 a.m. on Friday, January 2, the temperature here and in Republic was minus 40 degrees. The Case Street School let out at noon. Five boys at the Jackson School who didn't like a teacher tried to eject him from the school. Shooting revolvers at night in the city should be banned. There are a lot of scared people. And at the Breitung House, Madame Hewitt of Chicago will be telling fortunes all week. The price for sending a ten-word telegram has been cut in half to only 25 cents. The January 29th paper tells us that the temperature has been down to -44, and -47 in Republic. Many thermometers have broken or frozen up. People are asked to be aware of the poor and give aid, especially to those too proud to beg.

On the first week of February, the Maitlands gave a party for about 70 of their friends. Two drunks were picked up and put in jail. One said he was going to NY, and the other said he was going to H—ll. Congress is adjourning in March, but Mr. Breitung will go to Georgia until weather improves here. Mr. A. Gallant, of Republic, is charging his wife with an abortion, and says he has evidence. Mr. MacDonald's Hall has been so successful that he finds he needs to enlarge it by another 30 or 40 feet. It is also going to be used as a roller rink. The city has added over 30 gas lights in town. They are brilliant! A fierce dog in the Pierce block fastened his jaw on the leg of a ten year old Martin boy. He lost a lot of blood. The Breitung House, keeping up with the times, is going to install gas fixtures.

By March, things were not very good in Palmer as no mines are operating. Many of the residents have moved to Negaunee. The town physician, Dr. Haskel, has moved to Saginaw. The MH&O has purchased the M&W railroad and will adjust the tracks where needed. The trains will now go to Marquette via the MH&O and return on the M&W rails, the most gradual grade of the two. The Cambria Mine is putting in a new shaft by the side of Teal Lake and it will go

down 250 feet. Beware of counterfeit half-dollars with the year 1875 on them. March 19 and the temperature was 20 below this year. Under 15, there are more boys than girls. At over 90 the ladies outnumber men 4 to 1.

31. A Roller Rink Behind the Breitung House

A new building 80 x 100 feet will be built for roller skating behind the Breitung House on Lincoln Ave. according to the March 26, 1885, paper. It will have hardwood placed on edge over two-inch planking. Roller skating must be popular now as there is a rink operating also in the Pierce Building on west Iron St. The paper notes that there are many fine skaters among the young people. It is a good spring for mining here. The Saginaw Mining Co. has found good ore on the Maas farm, the Wheeling has a shaft down 80 feet, and the Argyle Mine has hired more men, the Wheat is running and the Milwaukee shipped 1000 tons this month. Chris Bollman, confined to bed for several months, sent his wife to the pharmacy and committed suicide. On Marquette St. a thief not only took clothes, but the clothesline, the reel, and the pins. Even Mayor Maitland is in trouble and may not get re-elected. He drove through someone's geese and killed some and then failed to stop. Turin town not only has a post office, but now has a column in the newspaper.

Quite a few stores are now being lit by gas. However, a disaster was averted at the Breitung when a gas valve was left on in a room. The gentleman was used to just blowing out the light of a kerosene lamp and did just that with the gas light, forgetting to turn off the valve. In May the ice rink was dismantled and the wood sold. Mr. Griffey doesn't think it made money this year. Mr. Tom Flynn has become half owner, with Mr. Griffey, of the <u>Iron Herald</u>. The Teal Lake Gold and Silver mine will build a shaft north of Teal Lake.

The Fire Department put out two fires that could have been very dangerous and are thankful for the new water system and the many hydrants with an endless amount of available water. More ladies are now being bitten by dogs on Jackson Street. The city will begin to eradicate all unlicensed dogs. A "filled to over-crowding" audience attended a play at the Methodist Episcopal Church when ten young girls did "Ten Virgins." Three stores in May put up nice awnings. Eighteen men are currently in the County Jail. Some could have been simply sent "out of town." Alfred Moberg is building a commodious house east of the Rolling Mill property. There is a scarcity of housing and of laborers, especially painters, masons, carpenters and mechanics. It is not so good at the Fayette Furnace where $500,000 of pig iron is unsold. Still lots of cows and horses on the streets. Driving around the Iron Cliff's Road (Drive) is becoming popular. The Negaunee Gas Light Co. has made the manufacturing plant more efficient and lowered the gas price by 20 percent.

The new roller rink is up and running and the name for it will be the "Adelphi." Special skaters will come from Wisconsin and perform. Lots of elegant fences are going up in Negaunee. $1,500 will go up in smoke on the evening of the July 4th celebration. Coles Circus came here on June 20th, with the world's largest elephant, "Samson." They lost money here as weather was bad, and they were charging 75 cents. Everyone is trying to get the new prison. Negaunee is trying, and Peter White offered to donate land for it in Marquette.

Michigan law now requires a test in order to dispense medicine. Those in business for over three years are exempt.

MacDonald's Opera House will be the largest in the U.P. The Grand Opening will give away prizes. A horse and buggy are the prizes. In August, bathing continues at Teal Lake and no one is certain whether it is legal or not. We find that Jack Carkeek, the wrestler from here, is now in the west and just beat the California champion. The Palmer Mine will sink two new shafts. The Union Depot has built a lunch counter. A new ordinance prevents bathing in Teal Lake. Marquette and Ishpeming will get postal delivery, but not Negaunee. There are now 1,657 children of school age. On the east end of Teal Lake, by the Water Works, 45 bushels of cranberries have been harvested. Several people witnessed a cat with a rat in its mouth that it could not kill, so the cat carried it over to a pail of water and held its head under until it drowned.

Hand drills in mines are being replaced by steam and compressed air driven drills. Another townsman has shot a large white owl. Philip Martin of Humboldt ran off on his wife and eight children in November. Water mains are now reaching to Peck and Tobin. 30 more people have signed up for water. Criminal Patrick Benan, who escaped from the Marquette Prison, and was caught in Escanaba, has now tried to escape from Jackson Prison, getting two feet from the top of the wall. Average school attendance has been 573, with over half neither tardy or absent. Girls are now getting their hair cut short like men. With Lake Superior frozen, Copper Country copper is coming through Negaunee by rail. People are reading a book by General Grant of his memories.

1886

Refrigerated railroad cars will make deliveries on the C&NW twice a week. In spite of four feet of snow in the woods, and that it is January, the Eagle Mills Lumber Co. has hired 200 workers. Others with horses can earn $5.00 if they are the first ones to hitch up to the hose cart when there is a fire. Several groups of people have seen a 7-foot man in a dark gown in the Jackson location that doesn't speak, but waves a hand. There are a lot of raffles, including for a mare, cow, stockings, and an accordion. Numbered tickets sell for a dollar, sometimes 50 cents. A snowplow was tried on Iron Street and did a good job. The city should get one. There is lots of snow. The original MH&O track through Carp Hill has six feet of snow in it and hasn't been kept open. The RR is using the Mqt and Western track it recently bought. Boys are jumping off of trestles and almost getting buried when they land.

There is a large demand for valentines. The habit is not dying out. Men got one from the Iron Cliffs in the form of a ten percent raise. Tobogganing is popular and the youth have built a hill on Teal Lake Bluff's south side.

32. Another New Railroad Being Surveyed.

A new railroad is being surveyed from Marquette to Negaunee. Mr. Griffey guesses that it might be the C&NW as it doesn't have a Lake Superior Dock. James Root was killed at the Pioneer Furnace when cut in two by a railroad car, but was able to say a few words before dying. Nearly 100 English, Finnish, Italian, and German men wish to become citizens. Sand has bubbled up in the bottom of the Negaunee Mine Shaft and rose quickly. All men managed to escape. The surface around the shaft sunk about 20 feet or more. There is a lot of exploring going on and many new ore bodies are being located. The Pioneer Furnace has much unsold pig iron, and may close temporarily.

The Breitung House is again making improvements. Now it will be steam heat to every room, and better yet, a call system to every room from the main office. A traveling museum, the Gilchrist, came to town for two days and for ten and fifteen cents it was well received. Average pay at the Jackson for March was $2.10 per day. The MacDonald Opera House closed up a show by Weber's Blonds. It had lots of vulgarity and filth.

33. The Coming of Bicycles

In June we find many people are purchasing bicycles that look like the ones we ride today. A club has been suggested. A railroad car, with an aquarium in it, is in town for viewing. People from Finland are now arriving here daily. The paper makes a plea for Negaunee to get a sewer system. The popularity of raffles continues with even homes being raffled. Try fishing for trout in Mud Lake by the Cliffs Drive. Some nice ones caught there in 1886. St. Paul's Church will soon be lit with city gas. Harry Neely, a boy, and one of many, lost his leg when he jumped on a ladder of a moving train and his foot swung under a wheel and was smashed to jelly.

On August 26 it was 97 in the shade. The DM&M RR has failed after being built to St. Ignace from Marquette, and will be sold to the highest bidder. Lots of horse racing is taking place at the Negaunee Driving Park. $1,000 in prizes are awarded some weeks. Many Negaunee men taking options on farms and properties to work the iron found there. In October, wrestler John Carkeck returned here and will wrestle the champion of the world in Marquette. M. A. Gibbs is making his house at

Main and Pioneer into the Queen Anne Style. More raffles for a clock, mare, and stove, and ticket holders get to go to a dance. Harry Neely, run over by a train, is up and around on crutches. A black man with a banjo is entertaining at area saloons. In Nevada, the great Comstock Lode seems to have run out of gold.

Mr. Jim Hogan believes his boy drowned in a Cleveland Iron Pit. He has to pay for pumping out the pit himself and so he is raffling off his silver watch to raise money. After pumping, there was no sign of the boy. Merchants are missing nice items this Christmas, due to shoplifters. The high schoolers now have a column in the Iron Herald, and wish everyone a Merry Christmas.

MacDonald Opera House

Photo from Irontown, June, 1983. Opera Upstairs.

—Take a small piece of paper and on it put figures, your age dropping months, weeks and days; multiply it by 2, then add the figures 3,768; add 12 to the result obtained, divide by 2, and then subtract from the result obtained the number of your years on earth and see if you do not obtain figures that you will not be likely to forget.

1887

Marriages are getting fancier, in churches rather than homes. More people are involved, and people are going on honeymoons. On Main Street, homes are now building out to the cemeteries. The county sheriff has made seven men in smaller towns deputies. Wolves are still plentiful in the woods and should be attacked with poison. The Negaunee Mine has spent $150,000 on shaft #2 to get through the quicksand, but now at 120 feet down, it has found ore. Uncle Tom's Cabin, a show that features a trick donkey, a pony, two brass bands, and 26 people, is coming to MacDonald's Hall.

34. The Jackson Mine is Sold.

In February, 1887, the Jackson mine changed hands when Capt. Sam Mitchell, the new head of the company, and some men from Cleveland, Ohio, purchased a controlling interest. Mitchell owns 2,400 shares, and they plan to continue paying the $25.00 dividend to shareholders. The mine hired 125 more men and now has over 400 employees. The Concentrating Co. has sold its latest ore product to a Wisconsin company. The paper finds that people want more news, but when it involves them they say, "Keep it out of the paper." The fire department has given two widows of firemen, $25.00 each. People are often carrying life insurance policies now and John Westman left his wife and three children $1,000 when killed by a timber falling on him at the Cambria. Some amateurs put on a fine play, "She Stoops to Conquer." Jenson and Williams will give the lady who saves the most wrappers from their famous soup an Elgin Gold watch.

Water from Teal Lake has been improved when an extension out further in the lake occurred in

March. All the mines are mining and selling ore. The Lucy is being pumped out for taking ore to surface.

35. Ed Breitung Dies

Hon. Ed Breitung died of pneumonia at his winter home in Georgia. He left here in good health on December 2. He died on March 3. He left a fortune estimated at seven million dollars, and had $56,000 in life insurance and much mining stock. He does not seem to have left a will.

Up to now the passengers in railroad cars have warmed themselves with the use of stoves. The railroads now believe there is a better way to heat the cars. At the Negaunee Red Front Store you now can get "fresh" Halibut and Red Snappers. Dr. Cyr is retiring from practicing medicine and surgery and is going to spend time with land and mining. The C&NW is having 700, 20-ton, iron ore cars built. On a sad note, one of the two new Bohemian ladies in town lost a baby, and received help for a casket. "Several saw them walking to the cemetery carrying the casket for the sexton to bury." The March 31 newspaper notes that it reached 24 below zero in Humboldt.

At the end of March, it was also noted that Capt. Mitchell will sink a shaft at the rear of Dr. Cyr's residence to facilitate hoisting from the #7 deposit. It is now April and five carloads of cows have been brought to town and sold for $60 to $100. each. A dam will be constructed across the Teal Lake Outlet, to be opened and closed as needed. The building of the new Cliff Office at Pioneer and Main is underway and will be practically fireproof. Wrestler Carkeck is going to Cornwall to wrestle the champion wrestler there.

There will be no circuses this summer notes the paper as the railroads are far too busy with ore trains to make room for circus trains. The MacDonald Opera House, however, is entertaining with a dog show one night and a comedy show another. A new bank has opened. It is the First National Bank and officers are Maitland, Maas, and Yates. The city is also building a new school, and the old Case school has been purchased for use as the Finnish Church. School will be held in the fall at a divided up Adelphi Rink. R. Hubbard used candy to get a four-year-old girl into a barn where he practically destroyed her. It is hoped she will live even though left in "insensible condition."

In spite of what the paper said, a Circus appeared here in July. It may have been small and may have come with wagons. It also seems that most of the forests are now cut for timber or charcoal as most of the furnaces have closed because of a lack of fuel. The Pioneer Stack No. 2 is running. Crimes these days include people tearing down posters as soon as they are put up, and a horseshoe found wedged in a track to derail a train. Carkeck won the Cornish wrestling match. The Concentrating Mill has been leased by Bice and McComber, to make brass and iron castings. M. Quinn's mammoth general store will be fitted with a "cash railway" carrying purchase money to and from a main office. Natives have appeared at an African seaport to report that "Stanley" is not dead, but alive and well (of Stanley and Livingston). 530 men are now on the Jackson Payroll, receiving $30,000 a month. People are purchasing the bonds sold to build the new school. They bear 4% interest. 50 men are at work there. Buildings still do not have numbers and directions are only given by location to other known stores. Capt. Broad has replaced Capt. Merry at the Jackson Mine. He has 14 children and is currently in Capt. Merry's home. A month later Tom Pellow was residing in it. If you want to know where the Merrys are, the Captain is now employed at an iron mine in Virginia. The MH&O RR is now the DSS&A and it is running nice Parlor Cars and going now to the Sault. The weather has been mild yet up to Christmas and bricks are being put on the new city school. Only the chimney left on the outside.

JACKSON MINE - NEGAUNEE, MICHIGAN

Above is the small mine locomotive leaving the series of pits connected by tunnels at the Jackson Mine. It is going north at the west end of the Mine where it transfers the Mine ore into C&NW cars, or wagons going to the Pioneer Furnace. Now permanently fenced off. (The U.P. Digitalization Center Collection.) On Left: Early Jackson miners working with hand tools in Pit #2. Neg. Historical Museum.

1888

The C&NW is building to Michigamme, and will parallel the DSS&A from Fan Bay on Lake Michigamme into town. A team of horses with a load of wood went through 15 inches of ice on Teal Lake and drowned. The large Iron Cliffs Store has been sold and will now be the Wells and Blake. With so many men working in the mines, deaths and injuries are occurring more frequently. Two men at the Jackson were killed by dynamite at pit # 5. There will be "Cornish Style Dancing" at the MacDonald Opera House with $50.00 in prizes. Many logging companies have been forced to stop work because of the deep snow. Mr. Boulson lost his building to fire and will now build a brick building on Main Street. The firemen have gotten dumbbells and other equipment to stay in shape. Mr. Marsell is selling his house at Teal Lake and Case for $3,500 and going to Tower, Minnesota. It looks like this summer will be slow for the mines as not much ore has been sold yet. March opened with several warm days, then very heavy snow. Deer are starving. The Scandinavian Sewing Group will provide the public an oyster supper at the Fire Hall. For some reason chimneys catch fire more in the spring than in winter, so people are warned to get them cleaned. The Methodist Church will give an oyster supper and entertainment at MacDonald's on Good Friday. Wm. Borlace of the Cornet Band has a new trombone. The Sons of St. George will have a large parade on April 23 and 2000 are expected to march. "Mrs. Tom Sharp, whose husband and children have been sick, will dispose of a cook stove by raffle in order to raise money for food and the house rent." The DSS&A has ordered 782 ore cars, 30 locomotives, 20 passenger coaches, six luggage and four mail-luggage cars. There is a real shortage of tenement housing and everything with a roof has people in it. The C&NW says it has arranged to bring the Barnum Circus here this summer. Sells Bros. is also coming.

People entering Negaunee on the DSS&A trains pass by the cemetery and are complaining of the dead animals, including horses, that are hauled to the east side of it to rot and smell. They should at least be buried. The Presbyterians in May are building a larger spire to hold a bell. Citizens would also like to see more gas lamps in town. The Clifford Troupe returned again and is entertaining at the Opera House. The answer is not 28 or 42, but what is the answer to this: "If a hen and a half lay an egg and a half in a day and a half, how many eggs will six hens lay in seven days?" Tramps again are appearing on our streets. It's a late winter. More snow on May 18th. The Sell's Bros. circus is playing at Republic before coming here. A new wrought iron fence will be put up around the Iron Cliffs new office building. It is June 1 and there are still snowbanks around. The iron market also looks discouraging.

36. Cement Walks

The first mention of the use of cement (concrete) is made in the May 31 issue of the Iron Herald. Case Street between Teal Lake Ave. and McKenzie is getting some. The Barnum Calliope and Advertising Car came to town. Al Lang has a new ironing machine at his Laundry and employs five Chinamen now. The army of tramps has been routed. They were a drinking, dirty, and saucy set said the paper. Some more snow fell on June 1. It's hard on soda fountains. The paper notes that we are to the longest day of the year with no warm days yet. The city continues to put in water mains and 20 more have hooked up. Negaunee got off to a bad baseball start, losing its first game to Champion 24 to 2. The next week it beat Republic, 21 to 11. The remains of Edward

Breitung have been moved from the Negaunee Cemetery to a Marquette family vault. He was also placed in an elegant casket. Miss Kate Sullivan is the new switchboard operator: "The right girl in the right place." The Jackson Iron Company is using some of its property to lay out lots for homes. Jack Carkeck is back home here and will play on the baseball team this summer.

Mines still seem to be operating fine. The Jackson paid $21,500 to 523 men and the Negaunee mine, $8,500 to 170 men. You can go to Mackinaw Island from here with an overnight included on the Island for $3.75 round trip on the DSS&A. A merry-go-round is in town and doing good business. The Barnum and Bailey Circus will have 69 railroad cars. $1.00 for tickets (under nine is .25).

37. A Sewer System

Mr. Griffey tells us in his August 9, 1888, paper that a system of complete sewerage in Negaunee is to be commenced at once. On the other hand, the Concentrating Works (It never did seem to have one correct name) is being dismantled and carried off, although the main building seems to be intact. Some people out east lost an awful lot of money. The building will sit for some years. Sewer bonds were available for sale on August 16th, and will pay 4.5 percent interest. The DSS&A has purchased a rotary snowplow. Two 90 degree days and then a week of frosts set in, in early September. It is estimated that 20 per cent of potatoes froze. Quite a bunch of snow is seen at Humboldt in the middle of September. People are selling their cows for the winter, but others are keeping them as milk is scarce.

Mines still seem to be operating at a high rate. Mr. Baptiste Castelletto died at the Cambria when the roof where he was working caved in on him. As usual, the mine was "exonerated." In October, there are several cases of chicken pox. The English Oak band, which will become very popular, will give a concert at the Opera House. The Cleveland Iron Mining Co. has given their men a raise in wages. The school gave all the children with no absences an afternoon off. A tramp kitten walked into the <u>Iron Herald</u> offices, took to the employees, and plans to stay. It was November. The Eagle Mills have installed an incandescent light plant with 20 lights. A new $5.00 counterfeit bill is in circulation and is an "almost perfect execution." It is Christmastime and there are many children who should be reported for not being in school. Mrs. Deloria on Silver Street shot two men on her property. Neither were seriously hurt.

Early bicycles had no brakes at first, and pedals may have continued to go around without stopping. Some were gear driven. They were popular and clubs were formed. 4/3/1896

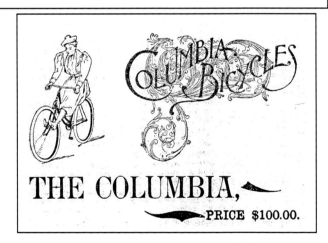

Two steam locomotives facing towards Ishpeming from the west end of the depot at the top right corner. Post Card

1889

There is talk of starting a boat club on Teal Lake. They would buy a steamer and some small yachts. Mr. Griffey notes that someone has made an invention of keeping milk from going sour and that perhaps it could apply to some of his readers. The city has purchased a snow roller and it will need 8 horses to pull it. It's leap year and several young ladies started it off by hitching up the carriage and picking up a boy for a midnight ride. There has been a lot of illness again, whooping cough, chickenpox, and mumps. 360 students of 400 at the new Case School have perfect attendance, however, for December. Teamsters now have the problem of boys jumping on their horses for a ride. On January 10, the <u>Iron Ore</u> came out with a historical 25 page paper. There is a lecture series and 200 tickets have been sold.

38. Electric Railroad Idea

The January 24th newspaper has a suggestion from a Mr. R. M. Adams from Detroit, suggesting the building of an electric RR from the Ishpeming Angeline Mine to the Negaunee Cemeteries on Main Street. Another suggestion from a local person is that when there is a late night fire, the whistle at the Pioneer Furnace should be used, and not the fire bell. No one heard the signal that there was a chimney fire last week. It was an interesting Masquerade Ball this year. Real Kickapoo Indians came to it from Milwaukee. It was packed. Negaunee will try for the next state institution. It will be a college.

March came and the city held a masquerade dog race. Over 3,000 came to town to see it. There was a husband and wife scrap out on Silver Street and the wife told the crowd that gathered not to interfere. The Pioneer Furnace is now installing its own electric light plant.

The city council may set aside some money for a reading room that has newspapers and magazines. The Dan Harringtons lost two children to diphtheria, ages 2 and 3, in two weeks. The Sam Collins lost their 14-year old daughter to it as well. The winter is very mild. There are raffles galore, including stoves, sewing machine, and the effects of a man who was single and died with no relatives here. That raffle raised $221. The money was sent to his mother in England. One could even win a horse, buggy, and harness for a buck. John Smith has built a set of steps from Mill Street to his home on top of Teal Lake Bluff. It's April and many notice the boys are walking to the town next door, and vice-versa. Waldman and Gripp have built many of the buildings and homes here, but have lost money lately and will go out of business. Over 300 men will lose their jobs. News in the April 11 paper noted that some loafers carried D. MacDonald's "Wooden Scotchman" from its place and threw it down on the street and caused some damage. It now can be seen at the Negaunee Historical Museum in 2009. Several residents who moved away were returned here for burial after they died. Such it was for Mr. Wm. Scanlon. The newspaper notes no large sales of iron ore yet. Each week there are write-ups of new babies being born with the names of the happy fathers. "Nothing more about an electric railroad," says the paper. It seems to have evaporated.

Have you seen saw logs in the woods? Spring came so quickly, they could not be taken out. It looks like Negaunee may not have a baseball team. The old are too old, and the young may be too lazy. Two mines have put teams together and will play each other for a $20.00 purse. A city team finally did get together and expect to lose their first game. Here is a puzzle from the paper: 1,1,1,3,3,3,5,5,5,7,7,7,9,9,9. Choose six numbers that add up to 21. A. F. Welin went to Sweden to spend the rest of his life. He is back.

Forepaugh's Circus is coming to town. That is why you see the kids picking up old copper boilers, old iron, and rags to get some pennies together. A large crowd gathered on the streets in late May to hear both the Salvation Army, and the English Oaks Band. The Yacht Club has ordered their steam boat and will build a dancing platform, dock, and waiting rooms for passengers.

The Street Railroad has received permission to lay track in Ishpeming. Negaunee should pass it also. It will cost $43,000. Local Scandinavian residents will underwrite the cost of bringing the Swedish National Concert Co. here. All of the Harry James family has come down with scarlet fever, and their 5-year old has died. Negaunee has formed a rugby team. It's called football. Here are the details of 1889 graduates. There are 6 girls and 5 boys. There will be a program at the MacDonald Opera House and the Negaunee Orchestra will play.

Captain George Mitchell died at age 54. He came here in 1864. Details in the June 6, 1899, Iron Herald. Sam Hooper did a somersault out of the back of a new barber chair in town. There will not be a repeat performance. Negaunee is raising money for the victims of the Johnstown, PA, flood. The Eliza Lobb steamer is now available for rides on Teal Lake for twenty-five cents. The Queen Mine now has the Concentrating Plant steam whistle. An ad is in the paper for dress-making at 212 Teal Lake Avenue. We don't know where the house number came from, but it may be a widow put it there, trying to make a living.

It's July. Surprisingly, the paper notes that tourists from all regions of the country are flocking here. A group of Cornish men from St. Austel, Cornwall, have arrived here to live. Quite a few homes are being built. Louis Kellan is the father of twin girls born on July 4th. And the nicest news of all is that more iron ore was shipped from the Lake Superior Mines in the last week than any time in history, and it is felt that a new shipping record will be made for the present year.

One of the homes being built is at the corner of Case and Teal Lake Avenue. A home that is there will be moved to the back of the lot and used as a barn. Moving older homes is now quite common. The Georgia Minstrels from Georgia are coming again. They entertained with black music to a full house. St. Paul's Catholic Church put a steel roof on, but it leaked so badly that several fine oil paintings were damaged and they have put on a new shingle roof again. The Pendill Mine Shaft directly in front of the Gold Street Brewery has found ore at the bottom. In August, "hundreds are out" picking huckleberries. And a very important vote passed, and that was to build a new city building.

Doctors here are suggesting the people clean the alleys since several cases of typhoid are again in existence. L. Corbett has purchased the 25 acre Negaunee Driving Park. He will keep the park, but sell lots for home on the property. The Iron Cliffs Co. will also now develop and sell lots from Clark Street toward Lake Bancroft. Two more bad injuries from trains: A Yelland boy lost fingers on one hand when a train could not stop fast enough to avoid killing him, Mrs. T. Shannon, walking on the tracks, died, after having both legs cut off by a train.

39. A Daring Stage Robber

There is an article in the paper of a stage robbery just five miles from Bessemer, Michigan. We put it in here as the story got more exciting when two weeks later, he was seen in a hotel in Republic and was taken alive. Redmond Holzhay had killed a man during the stage robbery, and terrorized the U.P. the previous five months. "His room was watched through the night and as he stepped from the hotel in the morning, Marshal Glode and E. E. Weiser each seized an arm, threw him to the walk and Constable Pat Whalen knocked him senseless with his billy, and the weapons ~ two revolvers and an ugly dirk—were taken from his belt." A full column records what he told of his several months of escapades. There was a little problem in Republic of who should get the $2,100 reward. (September 12, 1889 paper)

Two deaths here from typhoid. Teal Lake water is being tested. Newspapers around the country are stretching the story out of proportion. However, it is true that there are so many cases that Ishpeming doctors are now also involved with those ill. Another typhoid death in October. In October, Holzhay trird to escape from the jail in Bessemer. A ball and chain had been put on his leg and he used it to smash up his cell.

40. Plans for Union Park

On 10/24/1889, the first suggestion of building a race track and ball field between Ishpeming and Negaunee is made, with the possibility that the new Electric Street Railway might be also interested. There was a cave-in at the S. Buffalo and Queen mines of three acres. What happened on Halloween? Well, doorbells are rung, gates removed, and tic-tacs put on windows. Typhoid is still here. The Sam Matthews family has had five deaths. In November, Union Park is being built. Quite a few girls came on the train to visit Negaunee boys, but missed the train going home and went by "tie pass." (railroad ties) Tramps are being allowed to sleep "on the oak" at the Pioneer Furnace. Mr. Holzhay is now at the Marquette State Prison (11/28/1889).

41. Growth of Cave-Ins

There was a cave-in of ground between the Milwaukee and Grand Rapids Mines here. Two men present, but no deaths. However, on 11/21, a second cave-in at the Buffalo resulted in two deaths and 8 men seriously injured. A tremendous amount of air shot up the shaft, and six men were able to go to a safe spot, but suffered concussions. A large 200 x 300 foot area caved. A week later, the six are improving, two others are not. Note is made that "The necessity of a hospital in this community is becoming urgent." Other sickness in December places red cards in windows of homes having typhoid and scarlet fever. Also, there is a big problem of dogs now biting at the legs of horses and many teams going on runs. A new company, The Schlesinger, has purchased the South Buffalo Mine and will buy others as well in coming months. The new City Hall has gone up and the roof is now being put on for the New Year. Two drunken men wanted to pick up a drunken lady, but she had enough sense not to get in. It was a quiet Christmas, but the report of businessmen is that it was the best ever.

Ringling Bros. Shows at Union Park in 1895. Iron Herald, August 2, 1895

1890

Dr. Jones, the dentist, has purchased a "gasometer" to give vitalized air to patients during tooth extraction. People are noticing that the nice new city building, being built where the current one also is, has a dumpy old C&NW depot just below it. Several 5 and 10 dollar counterfeits are in circulation here. Mr. John Skinner is selling his "Creepers" which attach to shoes for frosty, icy weather. And here is another first: First National Bank is thinking of purchasing "Safety Boxes" for their new safe where customers can keep their valuable papers. The economy is excellent and even the Prince of Wales Mine is planning to hire 150 more men. The ore boats are getting better and bigger. The latest steamer is 38 feet wide, and 288 feet long, almost the length of a football field. Regarding Holzhay, the Gogebic sheriff claims he was trailing him just before his arrest, but the Wisconsin Central Railroad decided not to give him any of the reward. At the end of January, two young ladies had a real brawl, including "language that would make a fisherman blush." Charles Sundberg is going to build a handsome brick building on Iron Street.

People are saying that Holzhay is very quiet in prison and only speaks when spoken to. However, about March 13, Holzhay had himself a knife, and held the guard who delivered his food and got out of his cell for two hours but met Warden Tomkins, who then shot Holzhay in the arm and that event ended. The newspaper in Ashland complained that Holzhay need not have been shot.

Mr. Sundberg's new three story hotel on Iron Street will have two stores on the first floor, and a dining room and parlor on the second and rooms on the third. There is a mine behind the Breitung House on Lincoln Street with a shaft down 70 feet, and called the Elba Mine. We often wonder about the man who holds the drill for the men swinging the sledge hammers. The <u>Iron Herald</u> on April 10 tells us this "While turning a drill in the #9 pit of the Jackson, Sam Goodman had the second finger of his right hand crushed flat by a blow from a sledge. It was painful, but he will not suffer the loss of a crushed finger."

42. Jackson No. 7 Cave-in

There was a cave-in down inside the Jackson #7 Shaft with 40 men trapped at the bottom. Word spread quickly that they were without hope of recovery. Women who had come with their husbands' dinner pails waited in fear for word from down below. Finally one miner appeared at #8 and reported that some frozen ground and ice had to be removed from a drift, but the men should pass through to safety shortly. When the men did appear, they were badly frightened and trembled. It was lucky that Capt. Mitchell had completed a drift from #7 to #8 when he took over three years ago. The ore pillars had been removed from levels 1 and 2, which then caved in and caused the pipes on surface to break, as well as the large water tank to be crushed and the foundation of the engine house to tilt to the west, and the shaft to fill with dirt. (4/17/1890)

It is April and the paper notes that there are so may children on the streets that one would not think that school is in session. Another large powder and nitroglycerin explosion occurred at the Anthony Powder Works. They had just bought the old Gluckodyne Works. There were two large blasts and a fire. No one killed, but many windows broken again, even in Eagle Mills. The shock was felt in Marquette. Mr. Holzhay is now refusing to eat in prison. It is May now, and the first day was 11 degrees below freezing, or 21. The Oddfellows had a turnout of 645 persons for a fine meal at the Adelphi

Rink. If you want to support the Negaunee Baseball team, buy a share of stock for $10.00. 100 shares are being sold. A newspaper in Detroit notes that it would be dumb for Holzhay to starve himself, saying that many murderers are paroled after seven years. Two more small surface cavings at Jackson pit #7. A large safe has arrived at the depot addressed to August Carlson. No one here by that name, and no one has claimed the safe.

You are probably wondering after many pages, if there are still trout being caught in Teal Lake. In late May, Peter Larson, a young lad, caught one over three pounds. There are lots of homes going up on Snow, and Mill streets, including the Mission Church. Medor Gauthier will have the last home on Main Street. It is right next to the Catholic Cemetery.

The Concentrating Plant went up for auction in June. It did not work out well. Some items were sold, but not all of the machinery, and not the $40,000 building. There are a group of white men called the Order of Red Men, Tribe no. 22, and they held a parade and picnic at Teal Lake Grove. The paper reports that the Negaunee Baseball team lost, but plans to redeem themselves at their next game. No score was ever printed and everyone in the town must know what it was, but us. The DSS&A now has eight trains leaving the station here each day, with four in each direction. Plus there are other railroads. We also find that drinking bottled water is not new. Julius Vashaw has a large list of customers for his "mountain spring water," ten delivered quarts a month for $1.00. Adding up all the mines and the railroad wages, the newspaper arrives at a total monthly payroll now at $150,000 for just Negaunee residents. The Eliza Lobb steamboat is going again on Teal Lake. A group of 15 can go for a cost of only ten cents for an hour ride. It holds up to 25. The route of the electric railroad has been given for

Negaunee: "from Iron to Cyr, through the Jackson Location, then westerly along the old road to Ishpeming within a few steps of Union Park." Van Atta and Cook now have 12 tailors at work making coats and would hire more if they had room for them. The old city hall will be moved to Clark St. from Silver St. The Jackson House will be torn down for the new Sundberg Building.

Election laws have been changed and voting must be in secret, and ballots cannot be handed to people outside the building, nor taken out. Burglars tried to open the Jackson Co. vault and rob it of the next day's payroll money, but after drilling a hole in the vault, they could not open two safes. A new state law gives free textbooks for use by each student. Negaunee has a pretty women's baseball club with striped hats, kilt skirts, and black stockings, and are even proficient in the sport. In July, the Jackson Mine begins paying its employees with checks. Two carloads of cows were sold here a few months ago and the town now looks like a large barnyard. Woo Ah Len of the Chinese Laundry has hired two more men. Ah Chin, who sold his laundry three years ago and went to Minneapolis, has returned here and said, "Negaunee is goodee alle samee." The Negaunee Men's baseball team is at the bottom of six teams. In the Marquette Prison, Holzhay has found his girlfriend is now going with someone else. A week later he made an attempt on his life. Negaunee now has a Mr. L. S. Richards as a veterinarian and he will even do dentistry on horses. The Presbyterians have outgrown their church and are adding on 37 feet. The Schlesinger Mines (Queen, Prince of Wales, Buffalo, and S. Buffalo) now have lights at night from a Westinghouse Electric Plant.

The foundry building of the old Concentrating Plant has been purchased and will be made into ten 3-room apartments. Lots of deaths and injuries from accidents at mines and at railroads. L. Corbett spent $3,000 for his planting of 3000 bushels of potatoes. Now they are worth $8,000. By late October, there is 12 inches of snow on the ground and runners are in use. In November, Father Eis left St. Paul's after ten years and paying off the $14,000 debt. In the next week's paper, it was noted that it was the Bishop who paid off the debt. The St. Paul Railroad plans to build a track from Champion to Marquette. The DSS & A plans to build from Negaunee to Gladstone. Many other railroads are expanding (11/14).

The Jackson shaft #7 is open, but at an angle and does not look too safe. Not much other news in December, but lots of ads in the paper. One young fellow is going to Ishpeming seven nights a week, and is liable to commit matrimony. Christmas is spelled, "X-mas!" Two deaths recently are from typhoid. John Carkeck, the wrestler from Negaunee, will wrestle a Mr. King in Iron Mountain and says he will pin King down three times in one hour. And the year ends with the worse news of all: many men of many mines suddenly are being laid off because of lack of ore sales.

1891

Nice news to begin with. A gal came from Houghton to Negaunee to arrest John Zikoli with bastardy. He was arrested, married to the girl, and they left town rejoicing! Dr. Cyr has gone to Texas for a bit and is hunting some ducks – the same ones he hunted here this past fall. The Methodist Episcopal Church may build a new one as the old one is too small. Holzhay wrote a 14-page autobiography of his life while in prison. He sees his life as misspent.

Is there a problem with the water intake at Teal Lake? One customer had a 3 inch minnow come out of his faucet this week. The

Presbyterians are installing a new organ with 1000 pipes. Jack Carkeck got a carpet tack in his knee while wrestling and now has blood poisoning. A new mine, the Barasa, just past the cemeteries, will have to go through 160 feet of quicksand with its shaft. In February, George Paul was put in prison for beating his wife and making his children stand outside. The city water intake pipe has been moved 11 feet below the surface, but 10 feet above the bottom. Many lynx around and a 35 pounder was taken by Schweitzer's Mill. Mr. Carkeck's knee is evidently okay. The paper reports he won a match in Iowa. Katie Egan lost a hand and leg when age two by a C&NW train. The company will give her a job at age 17 so she can be self-supporting. The people in Palmer now have a M. E. Church. Two little girls appeared at a doorway on Case Street in April and said they were hungry. It was discovered that the mother had not eaten for two days and the case is being looked into. About 40 women have been doing work at home for Mr. Boulson, the tailor. Now he has sold his business.

43. No Ore Sales

We are now at the beginning of the shipping season in April, but the DSS&A had visited all the mines, and not even received one order to haul ore. The stagnation continued into May when the Lake Superior Co. in Ishpeming laid off 100 men. The newly elected city council is having a rough time with much disharmony. They are asked to do something with the disgraceful cemeteries which must be the worst in the state. Six or seven houses are going up on Cherry Street. A six-year old has died of scarlet fever. It was 89 degrees in early May, and a week later, two days of snow flurries. At the Adelphi Rink, walking races, heal to toe, are popular with races of 3, 5, 12 miles, and one of six days. A man from England walked 130 miles in 23 hours. There is lots of excitement.

44. Thoughts of a New Cemetery

The city is thinking of expanding the cemetery and a suggestion has been made of land on the north side of the road down Carp Hill.

45. A $10,000 YMCA

There is talk of a Young Men's Christian Association (YMCA) in town with a building to cost at least $10,000. A cement sidewalk will be built around the MacDonald Opera House. The Sporley family is seeking to hire a girl to do housework "and find a permanent place with a small family." Another railroad? Yes. The people of Manistique are planning to build a railroad from there to here. And the newspaper on May 22, 1891, has the news that the Cleveland-Cliffs Iron Co. has been formed by the Iron Cliffs and the Cleveland Iron Mining Companies. There is a nuisance grounds to bury animals on Baldwin Kiln Rd. that is being used, but no one seems to be burying. Captain Sam Mitchell has 225 people subscribing to the new YMCA, and he has personally pledged $2,500 cash.

The George Merry house will be removed to the rear top of the hill, along with one other home, so that Merry Street can connect with Iron and be straightened. The United States has a smaller debt than any other country of the world except Germany. It is June and there are six high school graduates this year, one fellow and five girls. Simon Hardware has built two 15 foot rowboats of galvanized steel for customers. It is the beginning of the use of a screen for viewing when the M. E. church will show an "Stereo-optican" program on "The effects of Alcohol." Ten cents admission. Other programs also. Negaunee had 50 applicants and hired three teachers. There were lots of disappointments. There are still not enough subscriptions in order to build the electric railway. It's July and the area farmers need water badly. They have also had two frosts. Lots of building going on with 24 homes going up. On Snow St., $3,000 is being spent to modernize the Jackson School with steam heat, and 29 rooms are being remodeled at the Breitung House. Businesses are returning to the cash basis again, and believe prices could be reduced 20 percent. A heavy rain finally on July 24th, but another bad frost followed, killing some potato tops. The L. H. Stanley's lost their six year-old son to scarlet fever. The Davis Block will be built at Silver and Iron. The electric RR is now being surveyed. A large new steam shovel has been built at the Buffalo Mine and many are worried that they will lose their jobs. It can load 350 tons an hour. The Edward Blakes lost a 4 year-old son to scarlet fever. He was isolated and the other four children could not be with him when he died. $100,000 is needed to build the Electric RR. Negaunee has subscribed $57,500, Ishpeming $25,500, and the balance by Marquette.

The Barasa Mine is still trying to put down their shaft, but encountering lots of water. Lizzie Johnson, 8, died of peritonitis and was the head of her class at school. Mine rock on Gold and Snow Streets is being used as fill. 40% iron works very well. September mining statistics show that half of the U.S. iron ore mined comes from Michigan's 82 iron mines and 17 blast furnaces. There are also six copper mines, nine coal mines and 97 salt companies. 100 men are now building the Electric RR grade. El. RR cars will soon arrive. Michigamme and Champion have nice news columns in the paper each week. A group of "bums" have been causing fights in several saloons and everyone is having trouble bringing things into control. Police officer Kemp has resigned. There was another large explosion at the Anthony Powder Company. MacDonald's lost a window and building shook.

Carkeek is busy wrestling and won two of three in the last match. Another three matches next week.

George Korten is unable to take care of his family of four children. They are living in squalor. His wife has a bad reputation, and left for four weeks, and he locked her out when she returned. The Iron Herald notes that "the children are notably bright and could be raised to lives of usefulness," and that the State Board of Charities should be notified. The Jackson Furnace at Fayette has been dismantled. There are lots of problems with the Iron Range and Huron Bay RR. at Champion. There is lots of typhoid illness, and non-payment of bills. The Electric Street Railroad is using concrete for the power house foundation. They hope to sell extra electricity. The Republic Sun newspaper office was destroyed by fire. Negaunee's water system now has over 700 customers. Tommy Bond went to Detroit in October, where he hopes to get help for his crippled leg which he got from a colored sock. The surgery went well. The Scientific American Magazine says that diphtheria can be cured if you grind up onions and bind them all around the throat up to the ears. Men are being killed each week in the mines, or badly injured..

Mrs. John Larson lost her purse, and while looking in all the stores she visited, a little girl came up to her with it. She had found it and used it to play store. All Mrs. Larson lost was her sleep. Still more raffles going on in November. John Dabb printed 50-cent tickets for a fine double barrel shotgun. (The newspaper must do okay printing all those tickets.) A man died who owed for 14 years of newspaper subscriptions. Just before the casket closed, the editor slipped in a thermometer, a palm leaf fan, and a recipe for making ice. "That's where they all go." The Street Car track is now about completed, end to end. It is December 4, and the streets are now mud and slush. Tommy Bond is still at Children's Hospital but can walk without crutches. The year is ending with most men working and the mines have been doing quite well. There is also thought that the two towns will be joined together with lots being sold along the Electric RR on the Brass Wire Property. Santa Claus will appear at some stores. The M. E. church will build a larger sanctuary for $12,000 and has raised half already.

1892

We start the year by discovering the meaning of such statements as "He can't drive one bit." A girl says that about her boyfriend. The bit is the part in the horse's mouth, meaning he can't even drive one horse. The young men of town gave a fine dance to the young girls as a "thank you" for the fine leap year party given for them. The Trolley wire is now up on the entire route. The costume party at MacDonald's will give out $60.00 in prizes. A game of polo will be played at the Adelphi Roller Rink this week. If the tracks can be cleaned, the Electric RR will be given its first trial. The city is now plowing sidewalks with a team. The city has cancelled its gas lights in favor of the new electric. Everyone wants to buy stock in the Electric RR. Negaunee gets a second newspaper in February, The Negaunee Independent.

The owners of the Maas, Lonstorf, and Mitchell Addition have deeded their block 16 to the city for a park and it has been accepted. Another masquerade party here and 4 carloads of people came on the streetcar. Tickets can be bought for 50 cents a dozen. The cars have nice upholstery and good heating. The cars have good power up grades with even over 50 people aboard. In February, John Medlin and Ole Swanson died in the mines. Lots of people are sending money home to Italy, Sweden, Canada, England and the Isle of Man at the post office. In March, three stores have electric arc lights. Judge Clark, a judge and manager of the Eagle Mills hotel, is returning to New York. Katie Buckley, 20, has died from consumption, the third Buckley girl to die. Druggist Gill had his father arrive from England to visit at age 67. It is the first time they have seen each other for 25 years. The Street Railroad has stopped business ~ a large snowstorm. Lake Angeline is being pumped out so mining can take place under it by three mining companies. There were five more deaths of children by measles, and one death of scarlet fever. Negaunee students are now learning shorthand and typo-writing. In April there is another death from measles. In the Negaunee area, over a dozen children "of a tender age" died by the end of the month. The Presbyterian sermon this week will be, "A Glass of Beer." The Electric Light Co. cannot sell any more light systems as they have sold all their power..

The Adelphi Rink closes and the baseball team is formed. The streetcars now run to the Ishpeming Hospital. Captain Sam Mitchell and his daughter, Anna, left for Cleveland this week to watch the launching of a ship that will bear his name. Two scholars were disobedient and have been suspended. It is sad to see the parents "are standing behind the two boys." A game of cricket will be played at Union Park. The Iron Herald is selling its steam engine and buying an electric motor for its presses. Mr. Henry Hayden was president of the Bank of Negaunee when it went under in the panic of '73. He left here and news is that he became the active governor of the Alaska Territory for a time. The Manistique and NW RR has a problem in Negaunee in trying to cross the Lonstorf addition to a depot on Teal Lake. Schools closed for Arbor Day so students could plant a tree. From Cherry Street to the Negaunee Racing Track, 44 homes have been built in the last two years. In May, the large wall on the south side of the Jackson school was completed. (The school is now done, the wall looks fine.) Two "monstrous" locomotives for the Iron Range and Huron Bay Railroad passed through here on their way to Champion and rails will now start to be laid to Huron Bay. The Champion newspaper column notes that they arrived there on the DSS&A tracks. Two residents died from railroad car wheels, Charles Kirkwood, and Nels Knudson. Many high schools have been accepted by the University, but not Negaunee, because of not offering of a language.

So many people are riding the electric railway that several new cars have arrived, called "trailers" and are towed by a regular car. The railway has also built a 60-foot platform at Union Park. The railway powerhouse is also being enlarged as the city wants more electricity. The Teal Lake Steamboat, Eliza, is now called "The Mermaid." There is also another boat called the "Jumbo." Both are filled on Sundays. A three-pound trout was caught near the waterworks in June. There seems to be a race on at the cemetery as bigger and heavier monuments are being installed. The John Johnson grave just got a $700 one.

46: The End of the Pioneer Furnace

The most important news of the summer was the headline: THE PIONEER FURNACE SOON TO BE DISMANTLED. The article fills two columns. No. 1 stack has failed, and the CCI will build a new furnace somewhere else. Businessman John Mitchell is in Finland and while gone, his business is closed by a foreclosure. Birds are busy building nests in the new street lights. They, and the eggs, have to be removed. 30 Arabians, persistent and saucy, are in town, begging and selling trinkets. Bill

Nye says he has a cow for sale: Raspberry color, 8 years old, leaves for two weeks in the spring, comes back with a calf, is one fourth short horn and ¾ hyena, and says, "I prefer to sell her to a non-resident." An armless woman and a blind man played an accordion and a hand organ and collected nickels and dimes. In those days everyone seems to have been able to work.

The Pioneer Furnace made its last cast of pig iron on Tuesday night, July 26th. A well-dressed man tried kissing several women on Main and Case Streets. The CCI is trying to find a place in the U.P. for its new furnace where they can obtain 350,000 bushels of charcoal each month. While the IR&HB was busy putting the 1300 foot, 60-foot deep cut through the Huron Mountains, the Champion mine was closing up. It would spell the end of the RR the following year. Quite a few of the unemployed in August left over several weeks to work on the Great Northern RR. Lots of others are traveling: Massachusetts, England, Cleveland, Niagara Falls, Denver, St. Louis, Seattle, and New York. The Morse brothers who left here have gone bankrupt in Minneapolis.

Deer Season from Sept. 25 to Oct. 25. No deer to be taken by dogs, light, pit, trap, or in water. Half of the stock of the Ropes Gold mine can be purchased because of the failure of owners to pay assessments. Dr. Cyr was attacked in his tent at Little Lake by a wolf. It brought down the tent, but he managed to get his Winchester and kill it. The hide measures 7'5". Mrs. Christ Johnson horse-whipped two girls who showed up to rob her apple tree. In September, the hugger was about again trying to get ladies to sit in his lap. A married one screamed and was let go. A gal called Crazy Kate has an income, but continues to beg and steal food. A new newspaper column about Redruth: It is the original name for Three Lakes. Quite a few people are building cottages there, traveling by train from here.

In October, Alfred Plow, a sober and industrious man with an ungovernable temper, beat his wife. In jail, his four children came to visit him and there were a lot of tears, including those of the presiding officer who had to leave and go out on the street. The IR & HB have decided to make a track to Ishpeming and by the end of the year, they brought one of their 115-ton locomotives into Champion, the biggest ever seen in the town. A roundhouse was planned to be built, but no more is ever heard until 1900.

1893

In the past few years, the C & NW has told the city of plans for a new depot in front of City Hall, with the moving of the horrible looking freight depot further north. However, in 1893, there are now thoughts of building a joint "Union" depot with the DSS & A. Bankruptcies are increasing as there are still not many mine orders for ore. There were 16 in 1892, and only 3 in the U.P. the year before. Spring came after a long winter when sleighs were used for 144 days straight. The first boat into Marquette had to get through 40 miles of ice. The Prince of Wales Mine has put in a double skip. The one, fully loaded coming up, will be partially balanced with one going down. In April everyone was buying one of the cows that came to town on the RR. They are fine looking ones this year.

Many people from Italy are now coming here to settle. 60 men are at work installing the new city sewer system. George MacDonald is in the bicycle business and a new bike has pneumatic tires. The bicycle club plans to have an evening lantern ride. It is Thursday, May 11, and the ice has just left Teal Lake. During the winter, the Jackson mine abandoned the famous #7 pit,

and it has refilled quickly with water, also causing the big spring on the south side of Merry Street to start flowing again. Winter prophets have forecasted a hot summer, so get out your fur coats. The next week, May 24, it snowed for a day and a half with six inches on the ground. In June, 11 young people graduated from high school.

47. The Panic of '93

Even worse than the weather is the news that: "Indications are that there will be considerable suffering in the next few months among the mining districts." The city also received a nice letter from Mr. E. C. Hungerford who supervised the building of the Pioneer Furnace and named the town, Negaunee. Two loose dogs fought for a half hour and about chewed each other up. Almost 100 people stopped to watch. Duncan Clark's Female Minstrels were in town with a rather rowdy show, yet you can't find one man who says he was there. In July there is a note that idle men are quite conspicuous on the streets. July 4 was busy in the morning, with a lot of people leaving town for other places, and the streets were empty the rest of the day. Mr. Braastad opened two of his mines but gave no pay except credit for purchases at his stores. Captain Peter Pascoe, who began at the Republic mine in 1863, has retired.

"So gradual has been the reduction on work forces, that in some respects it is not fully comprehended" stated the paper on July 21. Most mines have had lay-offs or closed up. The C&NW and the DSS&A have laid off many over the past two months. At this time, Gold Street is the main entrance to town from the south and for the St. George Festival, a large arch has been built over the street and their name will be spelled out in colored lights. The Joseph Winter family's horse got into the feed in the barn and died the next day from overeating. A laboring man, who always paid his bills, walked into a local store and carried out a bag of flour. The authorities went to his home, and saw his children eating out of the bag with spoons and water; they left and paid for the flour themselves.

At the Schlesinger Mine, all of the single workers have not been paid for some time, and an injunction has been filed to prevent them selling the mine until payment is made. Winter and Suess are going to a cash system only. The closed Pioneer Furnace still has 700 tons of pig iron unsold. In September, the Singer Sewing Machine office in town was closed. The paper also notes that rents are being reduced for the winter.

It is September 25th, and there is another snow storm, like it was November. It is just four months exactly since the last one on May 25. A serious freeze followed this week. Wm. Buzzo's daughter was shot in the stomach by a boy out hunting. He stopped to talk with him before having him arrested and realized that he was once a boy himself and simply gave him a lecture. The girl will be fine. It is now October and L. E. Chaussee rode his bike from here to Marquette in an hour and 19 minutes. There are another 15 to 20 cases of typhoid again. Emil Julahalla has died. 28 men died near Crystal Falls when the Michigamme River ran into a mine stope just 15 feet underground of the river. John Brown recently lost a son, and his daughter is very ill. He also lost his cow, and is now raffling off his horse to buy another cow. "Tickets are 50 cents and all who can, should buy one." Many people from town are still taking the train to the World's Fair, "The Columbian Exposition." The Ladies Aid Society is collecting old clothes and will repair them to give to the needy. By October, there are 100 empty homes when last year, notes Mr. Griffey, there wasn't one of them. As the year

closed there were four more deaths from typhoid.

Teal Lake & Power Station, Negaunee, Mich. m-715

Negaunee built its power plant on to the Water Works in 1897. Municipal Light had arrived! Photo from the Negaunee Historical Museum.

Trains will hereafter and until further notice leave Negaunee as follows:

DULUTH, SOUTH SHORE & ATLANTIC.

GOING EAST.

Boston Limited	6:37 A. M
Marquette Passenger	9:28 A. M
Detroit Express and Mail	12:25 P. M
Marquette passenger	5:44 P. M
Way Freight	4:40 P. M

GOING WEST.

Bessemer and Houghton Passenger	8:40 A. M
Republic Passenger	10:30 A. M
Houghton Express and Mail	3:45 P. M
Duluth Limited	6:50 P. M
Way Freight	6:37 A. M

C. & N. W. R. R.

GOING SOUTH.

Passenger No. 4	7:08 A. M
Mail and Express—No. 2	3:45 P M
Freight & Accommodation	7:34 A. M
Freight & Accommodation	3:20 P. M
Fast Freight	1:14 P, M

GOING NORTH.

Mail and Express—No. 1	12:25 P. M
Passenger No. 3	6:44 P. M
Freight & Accommodation	7:08 A. M
Freight & Accommodation	12:10 P. M
Fast Freight	12:15 P. M

1891 Train Schedule for the Negaunee Union Station. Photo from an ad in the Negaunee Iron Herald.

1894

It started out as a nice year for Miss Kate Fitzgerald, a housekeeper. She is one of seven relatives of a lady who left $1,370,000. in California. The Matthew Kratz's have lost three of their five children to typhoid, with a 12-year old daughter dying last Saturday. The Catholic Church is taking up a collection for them. The New Year's dog race saw 50 boys and their dogs participate. One won a fine sled and the others, cash, up to $5.00. Only 500 men are working in the mines, unlike the 2,000 a year ago. There was a big fire in Marquette, and the city Steamer did good work there, pumping water 3,100 feet from the Lake. The city of Negaunee will auction off stock piles of the Blue, Negaunee, and Jackson to pay the taxes they owe. Newberry has won the State Asylum and it is now finished and there are already 910 patients. A lot of adults joined the children in coasting down the Teal Lake Bluff by Pioneer Avenue: A lot of hilarity. The Street Railroad stopped running this winter when heavy snow came and they could not afford to hire enough men to shovel the track. It opened again in April. The wrestler from here, Jack Carkeck, finally won the title of World's Champion Wrestler, and has also finished college and become a lawyer in Milwaukee. Trout are being caught at Teal Lake. The new city sewer system is working so well, it is also draining the swamp at Teal Lake. In fashion: ladies wide-brimmed hats are being replaced with high and small rim ones. A deposit of nickel has been found near Spread Eagle, Wisconsin. There is a movement to have 24-hour clocks like the rest of the world. "Too much confusion by the 12 hour ones," many from Europe say. Not too many mine deaths now, only about one a month.

In early June there was a freeze and snow blew several times in the air. The city will have a grand July 4th celebration with visiting fire departments holding a hose race to see who can set up and spray water the fastest. There will also be a parade, sports, and dancing. The old grandstand at Union Park needs replacing and the new will be only half the size. The many cows in town are real professionals at night when they open gates. Bike races at Union Park have been added to the 4th celebration. And during the summer, for some poor reason considering the mining conditions, the miners in the west end of the U.P. are on strike.

Mr. Heyn, a store owner here for 20 years, is retiring and selling everything at great discount and may move to Milwaukee. Ishpeming and Negaunee saloon owners are playing each other in baseball. The same for the Negaunee and Marquette printers. The hugger has reappeared; hiding in dark places and then comes out and hugs and runs off. All the ladies now seem to want to walk after dark without escorts. The people in Republic may like to know that the population there is now 2,312. Negaunee is 5,861, down 217.

John Fish was holding his head, kneeling at a chair, after trying to commit suicide with a gun. His tongue was cut from its roots, a hole was in the windpipe, and he had a broken jaw, and lacerated flesh about his neck and was living when his family found him. He was 50 and a good citizen. The replacement for the Pioneer Furnace will be built at Gladstone by the CCI. Many men here are hoping to move to Gladstone and work at the new Furnace. Al Boyer hired 150 huckleberry pickers (blueberries) for several areas on the Dead River area and they picked 1000 bushels which were sent to outside markets. Tim Sheehan, 18, died in August when the roof of a drift fell on him and buried him for three days at the Lillie Mine. Young hoodlums were beastly drunk on Jackson Street at ten p.m. and disturbing the neighborhood. People are gathering to watch the Mexican jumping beans in a store window. Dennis Dowd killed a 60 lb. badger here.

Lots of ladies are wearing "bloomers" now, which are like slacks and riding men's bicycles. One man bought a pair of socks in which a lady had left her address and said she was looking for a husband. The letter he got back told him she had two children and had been married for 4 years.

Many men are leaving to find work elsewhere as winter begins to show itself. One is going to South Africa. There are 654 students enrolled in high school in September. Two basement rooms will receive floors and be used as classrooms. The school sent a telegram to Ohio and one to downstate for the two teachers needed.

The C&NW has renovated its old depot and put on a new roof. The planned new one has not been built. At the Ropes Gold Mine, yellow water has been found in the bottom of the shaft. Men think it might be from gold, and say that drinking it causes a loss of thirst for alcohol. This note for the old outdoorsman: A watch can be used as a compass. Point the hour hand toward the sun. South will be halfway between that hour hand and 12. The big cannons at Fort Mackinaw were removed in November. There is a new income tax for those who earn over $4,000. Several carloads of Christmas trees have passed through town on their way to Chicago. It is December 14, and there is no ice in the Marquette Harbor. Many women are in need of work this winter, so call on one of them if you need washing or ironing. The M. E. Church will do a play at the Opera House on Christmas Eve called "Santa Claus's Dilemma." Herman Thela, a cripple, is busy making his famous snowshoes. People like them better than the Canadian snowshoes. There are more ads mentioning Christmas, and some pictures of Santa Claus in the paper.

1895

48. A Mine Strike

The mines have been hiring again, and about 1000 are back to work. However, by summer many of the men have decided to go on strike. Republic and Champion miners have not participated. 600 men of the Lillie, Cambria, Negaunee, Jackson, and Davis mine formed processions 4 to 6 wide and marched in Negaunee and then in Ishpeming. The total area men numbered over 1,500. There has been no lawlessness, and no one seems to be opposed to the strike. Men are seeking $2.00 a day for contract mining, $1.75 a day for underground labor, and $1.50 a day for surface work.

Three bicyclists raced to Palmer and made it there in 16, 18, and 20 minutes. And heard on the street: "If bloomers have come to stay, I want to die." City police are now allowed to shoot unlicensed dogs, and there is a noticeable decrease in the number on the streets.

The strike continues on and the mining companies are finding the ore buyers are now going to other areas to get their ore. Still they only want to pay surface workers $1.25 an hour. In August, the large Ringling Bros. Circus came to Union Park. It took seven locomotives to pull all the double-size railroad cars. So many hundreds of people attended, and the street cars so full, that many walked to it. The strike went into its fifth week and people are concluding that it is a poor time to strike. Richard Cloke who left here in 1893 for Butte, Montana, is getting $4.00 a day as a mine timberman. The idle strikers are putting in their time by gathering in their winter wood supply. Eagle Mills has the Stickney Opera House and there was a grand dance there with music by Prof. Simkins. Two Negaunee homes were robbed without waking the residents, and 12 chickens stolen at another yard without a single "squawk."

Two Ishpeming mines tried to open up, but no one came to work. Several mines have decided to stop running their pumps, but Mr. Braastad had decided to pay the union wages at his Winthrop Mine. Steam shovels are running at the Buffalo, but it requires 57 guard troops to be there. Iron Mountain miners have sent $2,000 here to aid the miners' cause. Mayor Foley states that he is unhappy to have troops in town.

Women want to swim with their bloomer pants on. The men say "Okay, but stay in the water." Some miners in September attacked some innocent travelers who got off the train here because they thought that "Scabs" had come to town. They have been asked to be on their best behavior. In the tenth week of the strike, miners met at Union Park, and the vote was 700 to go back to work, and 600 to stay on strike. Men are going back to work, and the military is happy to go home, although some are dating local girls. Many mines that want to start shipping find that there are not enough boats to carry their ore. Also, some are handicapped in shipping by not having a steam shovel. Mr. Heyn finally retired from his business when Joyce and Mowie bought his Heyn Block and all his merchandise.

How is this for sportsmanship? The Marquette High School came to Negaunee for a game of football and lost 0 to 14. However, the Marquette team was treated to a fine banquet at the Opera House and then a night of entertainment with an orchestra. Marquette will invite Negaunee there next week. The Ishpeming Excelsior Furnace on South Pine Street is still running and now has a triple capacity engine and a third larger casting house. Woo ah Chin has returned from a trip to China to see his father, who died while he was there, and is now back at his laundry. In Marquette, the Negaunee football team won again, but was still royally entertained.

Mr. and Mrs. James Stanaway of Cornishtown now appear in the newspaper, rejoicing over a new baby girl. Jim will be opening up Sunday Schools all over the U.P. The CCI is grinding up bones for use as fertilizer. Dr. Hudson's name appears in the paper also for the first time when some girls stole his gate on Halloween, but wrecked it getting it off. They did more destruction than the boys this year. Nels Olson lost a valuable cow from choking on an apple. In her stomach they found 175 board nails, 11 pins, a can-opener, 3 iron staples, two pieces of glass, two feet of stove pipe wire, and more. Some think she died of an overloaded stomach. Hunters are allowed to get five deer this year.

49. The LS&I Railroad

In the November 22 Iron Herald it was announced that a new railroad will run from near Presque Isle in Marquette to Ishpeming and will begin construction immediately. Two weeks later, the name given to it was the Marquette and Iron Range. The shipping of ore has ended for the year and a new record has been made of 10,234,000 tons shipped on Lake Superior. Marquette County was down a bit this year because of the strike. A Belgian lady with three fine children has come to be a barber in Republic, and that is her name as well. And a wonderful thing happened at Teal Lake. Hundreds of skaters can go from end to end on glaring ice with no snow on it.

The C&NW "Negaunee" Summer, 1981 North W. Lines

1896

Great Wealth lies under the city of Negaunee says the Iron Herald, but the Fee Holders have the miners at their mercy. There are long newspaper columns about this subject, and for several weeks. It has to do with assessments and leases from land owners. The Milwaukee and St. Paul RR that goes to Champion is now planning to also come to Ishpeming. The Negaunee Mine is now electrified with power from the Electric Street Railway. The paper also says that men are busy building an extension of the railroad in Cornishtown. This must have been the new Jackson loading platform.

On January 24, the front page reads: "The Negaunee Iron Herald is at present waging war against the royalty sharks that derive enormous revenues from the mining properties in and about Negaunee. It has been a very nice winter and it's early February and the street cars are still running the regular route. Peter White, John M. Longyear, and Mayor Jacobs are forming a society to aid destitute children. In high school, Negaunee students are learning to write vertically, and it is claimed that it saves space. The Jackson Saloon gave a free concert and the big crowd that came proved quite remunerative. Mr. John Nesbitt has purchased a phonograph, our first record of one in Negaunee and is making a commercial profit with it. The DSS &A depot now has electric lights, but the C &NW depot still uses the old kerosene lanterns.

The Schlesinger Co. that has several mines here are now on trial in Milwaukee regarding its finances. The newspaper, in February, attacks those who rent mines, saying that Negaunee "has the greediest men to be found on the face of the earth." There was a bad accident at the Republic Mine when a skip holding 11 miners dumped. Four have died, three badly injured. One who died was one who escaped at the Mansfield mine when the Michigamme River broke into it. Two of the dead held $1000. life insurance policies. Mr. Maitland has taken a 3-5 week vacation to rest in the Bahama Islands. It is interesting to note that local people are often buying U. S. Savings bonds rather than putting their money in the banks here. E. C. Anthony attended an encampment meeting of Civil War soldiers downstate.

The Marquette and Iron Range RR (future LS&I) has purchased 11 engines, with 4 of them weighing 140,000 tons each. Ore cars have also been ordered. An interesting article in the June 12, 1896, Iron Herald notes that the railroad is thinking of ten cars with generators to make electricity during the 15 miles down to the dock, to be used to haul the empty cars back up the grade. No further word ever heard about this. The newspaper got it from the "Marine Record."

There is a lot of mine activity in April when the Lucy Mine began pumping out water. The Barasa Mine, by the cemeteries, will try putting down another shaft. The CCI is reducing forces as the stockpiles are full. The Queen hired 400 more miners. The paper notes that the Cleveland papers tell of little happening in the iron ore markets. Mr. J. S. Mitchell is selling 27 brands of bicycles and many men are learning to ride. It is interesting to note that guns still seem to be everywhere. Mr. R. G. Quinn was held up by three rough looking men on Case and Brown Streets. They wanted to get money and Mr. Quinn's diamond from his scarf. However, the store owner on the corner took out his gun and said: "This is what you will get." They quickly scattered.

The LS &I passenger depot will be at the head of Iron Street, south of the County Road. The new CCI Charcoal Furnace in Gladstone, the largest ever built, was lit. It replaced the Negaunee Pioneer Furnace. The Ishpeming Mines, including CCI, have let 750 men go. Louis Cyr's daughter is 8 years old and can lift 333 lbs. The Teal Lake "Mermaid" will now run on coal rather than gasoline and will be called: "The City of Negaunee." Seems like a step backwards. Some dealers are selling cigarettes to boys under 14. And a note to remind people that "wheels" (bicycles) have the same rights as horses. There was a big fire at L'Anse and many lost everything. The large Iron Ore dock downtown burned up, but has not been used for ten years. Mr. Thomas, the photographer, has opened a new business called a "Bicycle Livery" where you can rent, or buy, or get repairs done. We note that we have not yet seen the word "Bike" in the paper.

There are evangelistic special services from time to time, and Rev. Rowland has been at the Presbyterian Church for two weeks and the paper notes that "a number have accepted Christ as their Savior." High School graduates this year are five boys and 11 girls. Harry Trembath has been trying to ride a bike, but it throws him off and kicks him before he gets up. He pushed it up the street and may put it in a stable rather than a barn. Mr. Thomas in May had to order more bikes to rent at .25 an hour.

People saw a serious accident averted after a horse with a wagon jolted over some tracks to avoid an oncoming train, but a boy who had hitched a ride on the back suddenly fell off right on the tracks. A miracle happened when he quickly rolled over and was just missed by the wheels. There is a new paper in Republic again, called the Republic Record. Pastor Wilcox even has a new Columbia bicycle. It was stolen from his back shed. The Champion Hotel burned down, along with all the belongings of several teachers. A new hotel has now been built to completion and has 25 rooms. Samuel Gompers of the AF of L (American Federation of Labor) came to Negaunee to speak at the July 4th celebration. Stores were closed. The mine payrolls in Negaunee are now about $70,000 a month. Yet in Iron Mountain, idle men are living in boxcars, and 500 men have been discharged from mines in Ironwood.

A Horse and Canine Carnival came and the animals did everything but talk. There was a balloon ascension with it. So many people are buying bicycles that even the drug stores and a furniture store are selling them. Boys and girls may be using them to go out picking blueberries which they sell. The new railroad is now called the Lake Superior and Ishpeming (LS&I) and is building a large iron bridge at the Queen Mine. There is a medicine show by two comedians that the town loves, at Teal Lake every night. The M. E. Church had a band lead 300 children to Jackson Grove for a picnic. The city has set sizes and quality rules for all new city cement sidewalks. Wood ones are currently pretty bad and dangerous. Mrs. John Erickson was given medicine by the visiting Oregon Indian Medicine Co. for a tape-worm. When it came out it was 250 feet long. There was disaster for many miners when several mines made large layoffs in September. The newspaper estimates about half of them are now idle. There was another fire put out miraculously by the Negaunee Fire Department. A lighted lantern fell from a nail in a barn and a fire started. The owner got his horse out, and turned in the alarm. Noted the paper: "Firemen had the fire out with little damage except to some harnesses and hay." They were all treated handsomely afterwards.

Those wanting to be teachers can take the examinations tomorrow and those that pass will be given certificates. In October the city put its boxes over the fire hydrants. Only a miner or so are dying each month in the mines and some children as well, one from diphtheria. Mrs. Lavinia Jenkins Curtis died in childbirth at age 20 and leaves her husband and two children.

One nice new invention has been that the Electric RR has built a fine rotary snow blower to clear their tracks of snow. It is working so well that another town wants one made for them. In November, the Firemen's Annual Turkey Shoot was cancelled because of the weather, so the birds were "gambled away" by playing cards and won for about 60 cents each. Ed Harrington, who left for South Africa, has arrived, and tells others to stay away. He is diamond drilling. Tilden Township hired Essie Fisher as a teacher, but then hired someone else. The single lady sued and won, getting her whole year's salary.

Lots of babies are being born and the names of the parents are printed and congratulated! One man shot another large white owl and put it on display in a store window for Christmas. In the window of the Kirkwood and O'Donaghue Store, there is a miniature train that runs on a circle track using alcohol as a fuel. Two children died, one of diphtheria and another of scarlet fever. And the paper once again in the Christmas season notes that one look around to see who needs help, especially the sick and injured.

1896 Negaunee Firemen's Running Team for Firemen's Tournament "Hub & Hub" race at the "Soo." Iron Herald Photo, and from 1989 Negaunee Historical Museum Calendar.

1897

Mrs. Kronberg, who worked at a local lumber camp, was accidentally shot and her spine damaged. She has no feeling below her waist and is unable to leave the camp. Dr. Hudson went out and removed the bullet. Her husband is in the Marquette Prison until 1901 and she has an eight-year old daughter. The editor of the Iron Herald hopes that he will be given a pardon. In Milwaukee, Jack Carkeek, still retired from wrestling, bets that he can shoot more of 100 pigeons than Richard Merrill. He loses, with Merrill shooting 83, while Jack shot only 82. Of course, there was a purse of $200. and lots of spectators.

This kind note from the editor: "There i$ a little matter that our $ubscriber$ $eem to have forgotten." There is also a note that there are many counterfeit dollars in circulation and they seem to be worth the same. Popular Science says that the best way to thaw frozen pipes is to put lime around them and wet it. The chemical action gives off heat and will thaw the ice. In December, there is little snow and bicycles are out yet. Of course, there can be no dog and sled races this year.

50. A City Municipal Light Plant Arrives

The newspaper on January 15, 1897, notes that the materials for the new municipal light plant have arrived and are stored in the water works engine house building at Teal Lake. Electricity is coming into full use. Mr. John Mayne is building an "Apostle Clock" using Seth Thomas chimes and will have moving figures, and electric lights. Mr. John B. Orr is selling water filters that go on your faucet. "They work well." People are now playing a game called "Whist." Wm. Kirkpatrick has set up 8 tables at his home for players.   Several children are dying each month.   A Negaunee man got home very late and spent a lot of time quietly getting ready for bed. Just before he got in, his wife woke up and asked him what he was getting up so early for. To make his bluff good, he had to dress and go downtown.

Several men in Iron Mountain gave their own skin to help a girl there survive after her skin was scalded by hot water. She is recovering. The sisters of St. Joseph here netted over $100. in their performance of a drama called "The Deacon." Capt. Harvey, George Argall, and G. Reynolds are going to move to South Africa where they are promised fine mining jobs.   In spite of hard times, the Adelphi Roller Rink business is still attracting skaters.   Lots of weddings and most say they are in pastors' homes. The high school oratorical contest at the MacDonald Opera House was won by Miss Helen Reidy. In the same week she also won the inter-school contest in Ishpeming.   Her topic was "The Situation of the Negro."   She won $25.00.

51. Cinematoscope Arrives

The first mention of moving pictures is mentioned, still called the Cinematoscope. People had to go to Ishpeming to see it, but said it would have been worth twice the price.   It seemed to be a traveling exhibition.   Mrs. Uren lost her three-year old daughter this week, and Mr. and Mrs. B. Angelo lost two children, age seven, and an infant. Both died from diphtheria. The same week in March, a five-year old died from measles. Hon. John Jochim weighs almost 300 lbs and after destroying two bicycles last year he is having the Monarch Co. build him a special model called "The Jochim."   An all-lady Burlesque Co. will be here to entertain, arriving in their own railroad car.   A road is being completed to Goose Lake. The former

builder went bankrupt.   The Savings Bank has opened. No, it is not a bank, but a store, and it has hired three ladies to work there.   Still in March, the Robert Urens lost a one year old, and Policeman Miles Doyle and his wife lost their five-year old daughter.

The   Ishpeming   and   Negaunee   Athletic Associations will attend a large meet at Madison, WI, and have engaged a sleeping car and a dining car for themselves and friends who will make the trip down and back.   Jacob Dolf, the area Game Warden, had his report of violations and where he was on the days of February printed in the paper.   In April, Mr. P. B. Kirkwood beat Sam Mitchell to be the city mayor.   In Ishpeming, the Cinematoscope Exhibition returned a third time for three days. Not much happening in the mines this spring, but the Jackson reports sending ten cars of ore to Marquette every day. The mines have not needed much wood and so the Swanzy small-jobbers are also without work. Miss Reidy was now at the state oratorical contest, and placed 4th. She was offered a scholarship to Mt. Olivet College.

The Finnish are building their new Finn Hall. It is now up and enclosed. Quite a few people are adopting children from the Catholic Foundling Home in N.Y. A recent train here had 46 orphans on board, of which 4 or 5 came here. They all had tags on telling who their adoptive parents would be.   The fire department men and horses can be very proud of their work: Morgan Conway saw a fire in a building on West Iron St. and within two minutes of turning in the alarm at the box, water was going on the blaze from two hydrants.   Mail to Negaunee still reads "Negaunee (L.S.) Michigan." The L. S. stands for Lake Superior. (Oh, is that where they live?)   Another movie projector, invented by Tom Edison, will show here in a few days after being repaired. This one is called a

"Magniscope." A new law says you can't sell oleomargarine if it looks like butter. The newspaper printed Helen Reidy's oration in the paper, word for word. Chicago has sent orders to the U.P. for six tons of trailing arbutus.

52. The Fire-bug

By late April, the city is beginning to believe that there is a fire-bug at large. There were four fires in one week. In May there were two barn fires, then a fire of two barrels of paper, and a third fire that week as well. A fourth fire was an old building at the Pioneer Furnace. By the last week of May, at 10:30 at night, a fire was discovered well under way in the Brown Block Store. The fire department put it out but Mr. Brown lost some of his stock to the flames.

The Finnish have a drama group now, and they have been presenting plays. Their play this month was "Maantienvarrella" or "By the County Road." People picking arbutus are reminded not to pull out the plants by the roots to get the flowers. At age 71, Mr. J. B. Maas died, having come here in 1851. He ran a hardware store and then invested in several local mines, selling them at profits. With others, he began the First National Bank. He and his wife had nine children.

City statistics for 1896 were: 195 births (99M, 96F) 60 deaths (Only two from mine accidents). A bicyclist was given a ticket for riding his bicycle without a lamp on it. The paper noted that lamp sales will no doubt increase. The state has decided to have no deer season until 1899. The Savings Bank wants to hire a girl who speaks Finnish. As to the weather in June, there were three well-defined snowstorms, on June 1, 4, and June 6. (20 degrees at least one night.) More fires, with a torch being applied to burn hay. School is out in June and the teachers are leaving the area for Ohio, Wisconsin, and the Lower Peninsula. The bicyclists have rented Union park and racers can win up to $100.00 in prizes. Mrs. Kathrina Huhtanen committed suicide, leaving three children. John Mitchell will take the oldest, and one of each of the others will go to Ishpeming and Negaunee.

In July the waterworks plant is being expanded for the light plant. And even better news is that the Queen Mine has put 275 men to work as it reopens. Some boys fishing at Teal Lake found a dead newly-born baby. The Sault baseball team wanted to show how good they were so they picked a poorer team—mainly Negaunee to come and play them, but advertised them in the Sault as "the best team in the U.P." The Negaunee boys had their train travel, meals, and hotel, all paid. The game was so bad that after the Soo had over 40 runs and Negaunee had not reached first base yet, the game was called off in the 5th inning. Then the Sault group refused to pay the Negaunee bills. It came out okay. A note is also made in the newspaper that only recently have records been kept for the Protestant Negaunee Cemetery. Richard Searles has made a map and for a fee is taking care of some graves. Ex-Mayor Foley has a problem with deer that have shown up at Teal Lake to eat his nice cabbages. The state doesn't believe they ever trespass, and he can't use a gun to shoot them in the summer.

1,200 men are now working in the mines, the most in the last five years. 44 cows are wading in the water of Teal Lake to keep cool, right where the intake area is. The fires are getting worse each week. The dry goods store of Joyce and Mowrick had its interior engulfed with flames and smoke, but it was extinguished by the fire department. All goods were ruined. The fire department took the band to Ironwood. They gave our firemen two first prizes and six second prizes. Another fire. This one is at the railroad bridge over the Jackson Tunnel. Note is also made that a Finnish College will be built in Houghton. Obviously a change was made, as it ended up in Hancock.

The Steamer Christopher Columbus, a whaleback ship, will give a round-trip to Pictured Rocks from Marquette. Then it is scheduled to be cut up and retrofitted to take passengers to Alaska. In September, the city light plant is in operation with 34 arc lights. A barn fire was found at the MacKenzie residence. The Queen mine let go 175 workers. 30 or more Amish families have moved to the Newberry area. The Swedish Lutheran ladies held a bazaar. The handsome young ladies in native costume captured our town's young men and their pocketbooks. They cleared $350. Some Eastern Healers arrived in the early winter to meet with those who have health problems and were well received. They only saw ladies who were accompanied by their husbands, and children by parents. Some newspapers on microfilm were missing at year end. We checked the original copies, however. There was no one found who was setting the fires.

1898

Employees at local mines in the past year are listed as follows: Negaunee 257, Queen 254, Lillie 206, Cambria 162, and the Jackson 139. Other mines, probably also hiring Negaunee residents are: Lake Superior 720, Pittsburg and Lake Angeline 646, and the Cleveland Cliffs with several mines, 788. Adding in all mines of the Marquette range, the total men employed was 3,969.

There is a Copper Mine south of Marquette that has just shipped from its 40-foot shaft, twenty tons of rock to an Illinois smelter. We also start off the year with another fire of the Johnson Building on Gold St. Mrs. Richard Waters, a widow, miraculously escaped from the 2ⁿᵈ story.

The fire was set underneath the stairway. A second fire occurred at the Winter barn at Teal Lake Avenue the same week. There was a cave-in at the Cambria Mine when eight stopes collapsed on one another down to the ninth level. In Escanaba, the ore dock burned down, but it has given the Johnson Saw Mill here a lot of work cutting 150,000 board feet of timber for the new dock. Calumet is currently populated with 10,000 more people than Marquette.

The new Barasa shaft is down 168 feet and a drift in 55 feet, but no ore body has yet been reached. The closing of the Queen and the Jackson have caused two stores to close. The Swanzy Mine has been purchased and work is being done to reopen it. It was soon renamed, The Princeton. Of course, the name of the town changed as well. The telephone lines in Republic are going to be extended all the way to Pork City. D. MacDonald is putting an 18 x 31 foot dining hall and kitchen in the downstairs of the Opera House. "It will be commodious." said the paper. They are also purchasing a piano. The sale of valentines in town was quite large this year. A young couple came from Ishpeming on the streetcar, dressed in each other's clothes and were quite good, but people caught on anyway. Mr. and Mrs. John Bray left here nine years ago and purchased a farm in Kansas. They sold it, and now have returned, and the city has welcomed them with open arms. A buggy with harness, and a Clark organ, and the home of Eva Olmstead will all be raffled off with 50-cent raffle tickets. A new state law in Michigan allows judges to "marry" girls under 16 "who are disgraced." Negaunee's Dramatic Club will do a play, "A Cheerful Liar," and will use the money to support the City Band. It is interesting to note that the city currently has two of them, the Old City band, and the New. We don't know where the money went.

Some time ago we talked about those who rented mines and did not have to pay the taxes, or so they thought. The Michigan Supreme Court has ruled that the Pioneer-Arctic Iron Co. will now have to pay taxes to the city of Negaunee, which won the suit. Quite a few places put out their U. S. flags. It is now spring, and the birds, as well as the tramps, have appeared. Brook trout season has opened. Size is six inches, no limit, but can't be sold. Statistics for 1897 show that there were 45 deaths, with 29 being children under ten years of age. Andy Seass, manager of the Breitung House, has one of the old-fashioned bicycles with the large wheel in front, and small in back. So far, only one person has been able to stay on it. Bicycles can be seen on Negaunee Streets in the evening, and the town of Republic reports that 18 have been sold there this spring. In the national news, a U.S. Warship, the "Maine," has been blown up, and a war is on, but the newspaper can't get any news to print. The LS&I is the first railroad in the area to be purchasing steel cars for hauling ore, and it is also starting this summer to run passenger trains to Presque Isle.

Fires continue. A family left for a short trip to Escanaba and their house was set afire. Another fire was discovered. It was at the brewery on Gold Street. The statistics show that in the past year, Negaunee had 43 fires, and Ishpeming only had 15. The Cambria Mine is going to build a stable underground, heated by steam pipes, and use mules for tramming in drifts to the shaft. There are only candles for light, but they will have blankets and will be well fed. The Buffalo and Lucy mines are now working. Union Park is being rented this year by the Street Railway. Horse races are being held this year, and also a horse racing a bicycle (The horse won). A large July 4th celebration is planned, and after a boat race, a boat will be blown up to relive the moment that caused the current war to begin. Also, two city leaders will have a race. The day came of all the events, with $1,000 in prizes, but when it came time to blow up the "Maine," it failed. It seems that those in charge did not know how to handle explosives.

The Skerbeck family came to town and entertained with their circus. Quite a few people attended. John Grandberg, sentenced in 1891 for killing a man, has been liberated from the Marquette Prison after seven years. (Many murderers seem to serve short terms.) The city light plant continues to run well, and there are now over 500 incandescent lights in use. This is an improvement over noisy, smoky, arc lights. A new city law says that cows can't be on the streets at night, but it is not enforced. Maybe, notes the paper, what can be enforced is the cutting off of the cow bells.

Three men from Negaunee have now died in the war, the latest being Ed Mayotte. 17 men have signed up, but several smokers have been turned down for bad health. A few weeks later three Army men returned in poor health. Reports say that typhoid has killed more men than the battles. Then a fourth Negaunee man is killed. The demand for iron is increasing every week, but in spite of that, it is now official that the IR&HB railroad from Champion to Huron Bay is a failure. Negaunee people are exasperated regarding the fires of the past few years. The local police now seem to have a clue as to the firebug who has set the fires. An attempt was made to burn down the Opera House. Moving pictures of the war were being shown there this week.

We are going to read quite a bit now about Mrs. Arland from time to time. She owns a millinery store and in September is announcing that she is closing out her store with very low prices. Two Chinamen from Negaunee are going to China to visit their families. On their return, one had trouble getting back into the U.S. Several have made the trip in recent years so the laundry business must do well for them. A new railroad called the Escanaba and Lake Superior is building track from Lake Michigan to Republic. The state is planning to build a Normal School for teachers in Marquette County. The County Fair was held as usual in Marquette with 7,500 people in attendance, and the Negaunee Band gave a concert there. The Barasa mine continues to tunnel, and still has not found ore. They decided on using a diamond drill to help them find some.

53. A Golf Course

In October, it is announced that a golf club has been organized and a course set up by Union Park. (One can still drive by the remains.) The largest fire so far was set in the fall when two of the largest buildings on Iron Street and a home behind one on Jackson Street burned down. The fire department contained it using three streams of water from hydrants. The paper notes that a church bell takes about an hour to get an audience, but the fire bell only takes a minute. On a historical note, several old small wooden iron ore railroad cars were found by Republic, with the markings on them: "Bay de Noquet and Marquette R. R."

Saloons now have to be closed at 11:00 pm and on Sundays, and Mr. James Pendill, saloon owner, has been obeying. His saloon friends, however, in disagreement with him, tried to bomb his home last week, but bombed the wrong home. This week his home was blown up at 1:00 am and woke up the community, thinking it was the Powder Mill. Luckily he and his family were not hurt.

30 Gypsies are camping by the Queen Mine and are mostly reading fortunes. They come into town in fours and also do other things. About everyone in town is working, and to build a spur into the new Barasa Mine, laborers were brought in from Minnesota. Incendiaries still exist in town and at midnight on Sunday morning there was a fire at the Case Street School. The newspaper soon made a long column with the fires listed. That same week there was a fire in Mr. Muck's sausage room. The CCI is going to give sheep-raising a test. They will start with 250 of them kept this winter in one of the old Pioneer Furnace sheds. In a Christmas ad, the word "Santa Claus" is used. Prices this season are .22 for butter, 18 lbs of sugar for $2.00, bag of flour is $4.20, canned lobster is .25, and rolled oats, .04. The Street Railway employees all got a turkey, and no one in town was "locked up in the cooler."

THE SUNDBERG BLOCK

The Sundberg Block, built in 1890, with the Bijou Theater there in 1907 with the extra fancy front on the left. It was Negaunee's first permanent silent theater. The Liberty Theater (right) was across the street in the Hall of the International Odd Fellows. Both buildings still exist on the west end of Iron Street. Photo Credits: the March 1987 Newsletter of the Irontown Assoc., and Negaunee Museum.

H & L Wood Products
Original Liberty Theater Building

1899

Five hundred dollars will be the reward given to catch the incendiary person lighting fires for the past two years by the city of Negaunee. The same week there was a large fire of 60 tons of hay, and wagons, at a large barn at the Pioneer Furnace. At the high school, a speaker came from the university dental school to speak about the care of teeth. Mrs. Arland is selling a lot of her stock and will only have a millinery store now. Rosen Bros. of Milwaukee came to town and bought her merchandise for twenty cents on the dollar and will sell it for twenty-five cents. Then Rosen Bros. bought all of Mr. Quinn's goods to sell as well. The E&LS RR is now running trains to the Republic area. The Annual German Ball was held again and many came to see the costumes.

Fires continue. Ferdinand Winter's home by the Negaunee Brewery on Lincoln was set on fire. In addition, it has been very cold with many days below zero. Water pipes are freezing. There was an obituary in the paper for "Tom," the last original fire house horse. When the bell would sound, his stall was opened and he went right into the harness. It was noted that he was "powerful, willing, and intelligent." At the state level, there will be a Normal School built in our area, and every town is vying for it. In March, there is more snow and many deer are dying from wolves as the snow crust holds up the wolves' legs, but not the deer legs. A water pipe, down 7 feet, three inches, froze at the Queen Mine. Negaunee Senator Maitland is seeking to have a bill passed in Michigan that every fire will be inspected by a person hired for that job. The state of Massachusetts has such and has arrested 80 incendiary people. Rosen Bros. have purchased the stock of two more stores in Negaunee.

54. "Crazy Kate"

Crazy Kate is again roaming the city and entering homes and scaring residents. We'll hear more about her yet. Mr. Benj. C. Neely is the new city Mayor. Two more bodies of Spanish-American War soldiers have arrived for burial and there will be a parade with both of the city bands participating. The war officially ended in the past December. In April, the Carnegie Steel Co. became the new owner of the Queen, Blue, Buffalo, and Prince of Wales Mines.

55. Miss Philomena Cody

It was a surprise to the author to read in the April 21, 1899, copy of the paper that "Miss Philomena Cody, about age 20, is violently insane, and has been sent to Newberry." The Negaunee Historical Museum has a sketch and information about her. Now that railroad rails are becoming steel, the local railroads are getting 80-ton engines and 40-ton steel iron ore cars. They can get through the snow without plowing or snow-blowing. Rosen Brothers are having such fine business here that they have decided to become a permanent part of Negaunee. Hurley, Wisconsin, also makes the paper by having 48 taverns and 31 houses of prostitution including children participants. The town only has 1,500 people and only three butcher shops and two drugstores.

The CCI is now starting its new farm half way to Palmer and will house their sheep there this summer. A new type of entertainment came to town when Backman's Glass Blowers gave all attenders a glass-blown object, with children getting a glass boat. Over 2,000 tickets were sold. Mr. Jack Carkeek is still wrestling and will have a bout in Houghton. The local Presbyterian Church seems to be without a pastor now, and for some time. At the M. E. Church, Sam Mitchell has put in a contribution towards a new church to replace the present one, now 30 years old. It will be moved to the west and be joined to the new sanctuary for extra seating when needed. Those working for the railroads have been getting raises lately with the C&NW giving all a ten-percent increase. Mrs. Arland says she is retiring and having a sale of all items in her new store. A play came to town and was called "The Gay Matinee Girl." Now it's an obsolete title.

Wm. Fred Rice died at the Negaunee Mine when he fell 60 feet down the shaft and leaves a widow and nine children. The ninth was born the evening of the funeral. Mr. Richard Brown of Michigamme is proudly displaying a large beautiful loon and a blue crane he shot. The St. John's church had a big celebration. Special trains and street cars, 1000 dinners were sold, and five bands played. The DSS &A now has platforms between cars to walk from one to the other, even with the train moving. The men at the local powder mill knew how to get rid of a dead horse. They blew it up and put what parts were left in the hole that was created. A new attraction will bring a lot of men to the audience. It will be a bicycle race by ladies.

F. W. Read has purchased the Johnson Lumber Company and will close the operation. There is no more wood left to saw in the area. In July there is a lot going on at the Union Park. Carkeek wrestled and beat three other men, and three ladies held a one mile bicycle race. There were two other races. The Ringling Bros. Circus came to Union Park as well, and the street cars carried over 2,000 people before ten in the morning. Another circus came, and it is on movie film being shown to a crowd each night at the Winter and Suess building. The Negaunee Mine paid a 14% dividend on its common stock, and 7% on the preferred. At the saloons, a man and a girl have been giving all a real treat. He is a "musical prodigy" and plays all instruments.

His wife plays a guitar and a mandolin. The Normal School for Teachers will be in Marquette and it will open on Sept. 19. The county fair was done in, losing $1,500. due to a snowstorm. At the high school, the physics class hung a pendulum up 25 feet high and let it swing to show how the earth turns underneath. In them, there have been many deaths in the past few years, including Capt. Kinsman at the Champion Mine in October. Several men died by falling down shafts when riding in skips which jerked or tipped accidentally.

Negaunee is building a new school on the east side of town. Fire again. This time the shaft house at the Negaunee was lit on fire. The closest hydrant was 2,000 feet away. Captain Mitchell has decided to resign at the mine but stay on until a successor is found. Those that bet on copper stocks have also gotten burned. Prices are way down and Paine, Webber is closing here after "its patrons have been skinned." The old Concentrating Works building is being torn down by the Jackson Mining Co. for a temporary shaft house at the Negaunee Mine. The water from Teal Lake has been terrible as the intake fell off the cribbing and is lying in the mud. More fires: A large one averted by the fire department at the Jackson Mine coal sheds, and another barn at the Maloney Residents on Case Street was burned. Football is on and the new high school team played former players. The paper noted it was disastrous for the new team and no score was given. Mrs. Arland is still in business and is having a hat sale, but is leaving the area for the winter. Downtown, the eyes of the youngsters are bulging out as they look at all the items in the many store windows for what Santa might bring.

Before leaving 1899, we once again come to the name of "Crazy Kate." Her real name is Elizabeth Reynolds, and she is being sent to

Newberry. She had to turn over all her money for safekeeping and it turns out she had $1,290 in cash and eleven $100 bonds. As she needed some better clothes, the sheriff sent her home, and when she returned quite nicely dressed, she turned over an additional $2,455. She had been married for 37 years at one time and the sheriff noted that "dressed up she was not a bad looking woman." She had no objections of going to Newberry.

Here is a photo of the present road from the Jackson Mine to the old Wawonowin Golf Course and west. The Electric Railway from Negaunee to Ishpeming follows this road on the north side. As this photo is looking east, the Railway is on the left.

Here is a drawing by Sharon Johnson Fosmo of Philomena Cody. Although sent to the Newbery State Mental Hospital at age 20, in 1899, she was well known in Negaunee in later years. The sketch is at the Negaunee Historical Museum.

1900

Meat market, pharmacy, and post office of the Champion Mine Co-op went up in flames in Champion, but all the rest of the town was saved. The year starts off in Negaunee with a several week discussion regarding appraisals in Michigan for taxation. There are many inequalities of assessments. Here they are one-third of the value, but there are no state rules. A dividend has been declared by the Street Railway. It is the first ever and will be 4%, half payable now and half in mid-summer. They also have paid off their $50,000 debt. The newspaper has heard that there have been some cock-fights in town, but no one seems to have been there. The new Park School has opened with 250 students. The January 26th paper notes that warm weather has caused wagons to be back in use.

56. "Rag-Time" Music

As entertainment goes, Rag Time music is dominant at the start of the "Gay "90's". It must be on record players as well. The Alexander Maitlands gave a large "coontown party" with a prize for the most elegant costume. It was noted that music for the night was coontown music. At the MacDonald Opera House, 25 rag-time stars will provide a program called the "Big Coontown 400." On the program is also the new popular "cake walk."

The present Negaunee Mines operating are the Jackson, Negaunee, the Queen Group, the Cambria and the Lillie. Mines that will operate this summer will be the Blue, Barasa, and the Hartford. The Negaunee Mine's new shaft has made it through a large body of quicksand and is now going through rock. Single ladies of town held an "Old Maids Convention," and raised $131. for the Sisters of St. Joseph. Thirty black singers came and were excellent. They called themselves "The Hottest Coons in Dixie."

The new Methodist Episcopal Church was dedicated. Cheap Joe's Store has a new sign in large gold letters. Elections took place in April and the city mines closed for it.

Many people from Finland are arriving. Over 50 were on one train this week. Some got off here and at Ishpeming, but most went on to the Copper Country. Sundberg's Store has a nice ad for bicycles for $25.00. They claim they are $35.00 at other places. How are other prices? We see children's shoes for .59, ladies' shoes for $1.48 to $2.98, men's suits for $7.77, men's shirts for .48 and ladies' skirts and suits for $4.86 to $11.98. Ladies' dresses go to the floor. Jack Carkeek wrestled the champion of England in England, and lost two of three matches.

57. Negaunee Mine Cave-in

At the Negaunee Mine, Capt. Piper is now in charge. Levels 4, 5, and 6 caved in, and crushed together. Water has also now come pouring in from the old sand shaft that is on an incline from the vertical shaft. It has caved in as well and the skip is pinned in the wreckage. Caving at the mine continues. A week or so later there is fear that the entire mine might come down. The main shaft seems to be still quite good and pumps are busy trying to keep up with the water at the bottom of the shaft, amounting to 2,500 gallons a minute. The surface continues to move 100 feet east of the change house and 4000 feet north of the engine house. There are cracks in the foundations. The workforce of 503 men is idle and awaiting further word. Some went down with permission and rescued the eight dedicated mules who were already up to their bellies in water and very weak. They are now enjoying a warm sunny pasture. The mine hopes to begin taking ore from the top two levels and is still sinking the new sand shaft.

Mrs. Arland is opening a millinery store on May 21 and 22. More Finnish people passed through, and one group from here will go down state, hoping to buy a large area and divide it up for farming. Mrs. Sam Mitchell wishes to hire a cook. Mr. Maas wants the Catholic cemetery to lease him their mineral rights. He says the ore is down over 500 feet there. John Nesbitt, in his saloon, has installed a banjo machine. You put in your money, and the machine does the rest. At MacDonald's, the play is "The Runaway Wife," and ladies are allowed to attend free the first evening.

In Humboldt, the public school will graduate three eighth graders this year. Wrestler Jack Carkeet wrestled for the fourth time in England and won a Gold and Silver belt. The Civil War Veterans met in Negaunee and reenacted a battle on Gold Street, using 3,000 rounds of ammunition. In Palmer, the son of Dr. Moll disappeared. Eventually after much town searching, it was thought that maybe he went to see his grandma in Negaunee. The father found his son, with his dog, still walking and was almost to the Pioneer Furnace after going five miles.

How many French Canadians are in the U.P.? 3,000 of them met for a reunion in Marquette. In July, the Ringling Bros. Circus was here. There will be 1000 people, 500 horses, 65 railroad cars, 25 elephants, 100 cages, and a free street parade. There are 135 more horses in town as five car-loads of them arrived and are for sale at $50.00 each. Many people must now be able to afford a horse. News about chickens also: The U. S. Supreme court has ruled that chickens running loose are the same as wild animals and are offered no protection under the law.

In July, the Cleveland Cliffs Iron Company celebrated its 75th anniversary at its first mining area in Cleveland at Bluff Street and the Cliffs Cottage. It was noted that the Iron Cliffs Company, which was part of the merger, was invited, but rather neglected. Peter White gave a long and enjoyable historical address with lots of humor. Miners with over 20 years were given a gold medal.

I. E. Swift and company have purchased an "Automobile Wagon" for deliveries, and it is gathering a lot of attention on the streets. No one has thought of the name "truck" yet.
In August, the paper notes that some men at the Negaunee Mine have gone back to work and things look stable with the water under control. At the Newberry State Hospital, "Crazy Kate" has recovered her health and mind enough to be released, stating only that she is "eccentric." Her brother will use the interest of her $5,000 of savings to provide for her on his farm in Ohio.

There is an interesting note regarding the dismantling of the IR&HB RR in Champion. As the large rock cut through the Huron Mountain Summit has filled up with considerable dirt, the rail purchasers are laying ties and rails over the dirt in order to remove all the rails south to Champion. Mr. Vashaw lost his two fine horses and is now raffling off his gold watch so that he can purchase another horse. The State University has tested Teal Lake water and finds no spreadable germs. They suggest a good fence to keep cows out of the lake, and a sewer control for homes on the south side. A column on Turin tells us it has a Methodist Church, a large store, and a school, and another active religious group. In the fall, serious illness again appears. Park School let out because of four cases of diphtheria, and there are six cases of typhoid. At Pictured Rocks, a storm has destroyed the 100 foot-high "Grand Portal" which rose from the water line.

In October, the Negaunee Mine reopened with 325 men. It had 527 at work when the large cave-in occurred. In the same month, the Jackson Stump, left from the tree which stood on the shiny ore bluff was burned up. People are wondering if "Kate" will return to her home. It is quite badly destroyed by people who went in looking for money. The paper notes that it is only fit for a torch. However, she became too much for her brother and eventually the home was fixed up and she returned here. The CCI has purchased the first "Separator" at its farm and can now separate the cream from the milk of 25 cows, and make and sell butter.

58. Carbide Lamps Invented.

A new miner's light has been invented to replace the candle. It runs for five hours on one fill of carbide on which water drips and causes acetylene gas to form and be a light. The year ended with the death of the school superintendent, H. B. Grogman, 30, who died from typhoid.

1901

Patrick Benan, 37, who we read about in 1885 for killing the assistant sheriff has been pardoned and freed by the governor. He says that it was alcohol that caused it. At this time Marquette County has 127 saloons, with 25 in Negaunee. The paper also notes that there are a lot of "sparklers" around. It seems a lot of ladies got them as Christmas gifts. Statistics tell us that in 1900 Negaunee had 129 deaths with 57 of them being children under five. At the Negaunee Mine, work has again stopped when a drill found a large water "tank" underground and it has filled the bottom of the mine to 15 feet and buried the pump. A week later, the water had lessened to only 1,300 gallons a minute.

59. The Maas Mine

Mr. George Maas reappears in the news when it is found that he has purchased mineral rights from eight property owners, including the city of Negaunee and the Gaffney addition and plans a large mine. He currently is drilling with five diamond drills on Main Street and the Baldwin Kiln Road (old). People are putting up with the noise and smoke from the steam engines. A land rush has resulted in town. Ben Neely has taken an option of the East Jackson Mine and many others find that they have built over ore and own mineral rights. The Lobb estate had been considered worthless, being mostly a large swamp by the water works, but at a February auction, the bidding went wild. CCI bought the surface and half the mineral rights for $8,000.

The Oliver Company will lease the Barasa Mine property. The present owners feel that the day of small mines is quickly passing. The Oliver Co. also bought Mrs. Henry McComber's home between the water works and the old sawmill for $5,000 cash. The CCI by March had spent $329,000 for land purchases, and many of the citizens are becoming wealthy and independent. Offers have been so good that there are no condemnation proceedings. The city is wondering if it will still have a water supply at Teal Lake once all the mining begins.

The second part of the Street Railway dividends was paid of 50 cents per share. The State has declared that the Pioneer Iron Co. is no longer a reality and that the property legally returns to Edward Breitung as inheritor. Good Friday was celebrated with the English Oak Lodge Sons of St. George giving a supper at the Opera House and the City Band will give a concert. We also find out that the Milwaukee and St. Paul RR that goes to Champion has an ore dock of 150 chutes at Wells. Bicycles with chains are $18.00 and those with drive-shafts are $80.00 at Mr. Sjoholm's. George Sellwood is stripping the surface at the old Rolling Mill Mine, now called the Chester. Rosen Bros. store has a contest to guess the name of the doll in its store window. The doll holds an envelope in its hand with the name in it. High school graduates this year total ten. Turin is to get a chair factory. Eldridge Barabe is building a very large chicken ranch. The Street Railway is building a branch into the golf clubhouse near Cleveland Park.

At the Negaunee Mine, water that once came in at 3,000 gallons a minute has been reduced to 900 by diverting some to the old shaft for pumping up to surface. It is noted that the level of Teal Lake has remained steady. "Crazy Kate" is now back in town and the newspaper notes that they hope she will "conduct herself along the lines of decency." The C&NW passenger train has added on a dining car. One can eat supper going to Escanaba, and eat breakfast between there and here on return trips. Several persons have lost money on bad stocks, including the U.P. Adventure Copper Mine, and out-of-town investments. Mr. L. E. Chausee had bought 1000 shares of one defunct company.

It is reported in the July 12th paper that at three o'clock on Saturday afternoon, July 6th, that a large area of ground fell down 100 feet and that was larger than expected, it being a worked out area. There were 300 men back at work, but in a safe area near the new shaft. The cave-in took the foundation NE corner of the engine house and in one week Captain Piper had the engine and huge cable drums skidded 100 feet away to safe ground. It was no sooner completed and another cave-in came, taking the entire engine house and foundation and swallowed it up in the hole, including over 100,000 bricks. An even older engine house also hangs over the hole, which is now covering over an acre. The Iron Herald noted to the public that "there is no cause for alarm."

John Shea has purchased a 33-foot lot and will build a new two-story building between Iron and Jackson Streets on it. On a nice day in July, John Powers brought a decorated trolley car full of children to Negaunee to enjoy the Merry-go-round here and to have some ice cream at Perkins Drug Store. Dr. Charles Yemans, who was the Methodist's fourth pastor here, is now a successful doctor in Detroit, and even ran for Mayor. We will also hear more about the cows. One recently opened a gate, ate up the garden, trampled flowers, and left its card on the back step---and left the gate open. Perhaps the city should be designated at night as a "Cow Park." At the Huron Mountain Club, there is a need for two waitresses. Thomas Marcotte, 43, has died after suffering a stroke. He leaves a wife and five helpless children. In August, the Barabes begin building a fine new home at Brown and Main. (Now the Negaunee Historical Society Museum.)

Caving continues at the Negaunee Mine. It caused a compressed air pipe to break and the large Prescott pump stopped and is buried in the water. A new DSS&A RR track is being laid from the cemetery area to avoid the cave-in area. Local U.P. copper stocks of the three large Copper Country mines are doing well, selling from $170. to $710. a share. Smoke came from the roof of the Breitung Hotel, but the fire department quickly found out it was coming up through the walls, and found the actual fire in the basement laundry room in time to miraculously save the hotel.

The pure water supply at Iron and Silver Streets has some beautification going on as it is much used and better than that of Teal Lake. The Negaunee Co-op, a business purchased from

Mr. T. M. Wells, has paid its first dividend of 7%. And, another note now about "Kate." Her guardian asks that the newspaper inform people not to give her anything as she is not in need, and it only encourages her to do more of that practice. In September the paper notes also that things in Negaunee are quite prosperous and there is hardly a vacant house and all who wish to work have a job.
President McKinley was shot in Buffalo, and the MacDonald Opera House cancelled a large ball in memory of him. Another long-time store, the Savings Bank, is closing, owned by Mr. Davidson who is in poor health. The store has a large sale ad.

The CCI will put in a shaft 800 feet east of the Baldwin Kiln Rd (old), but will have quicksand for 118 to 263 feet down. This will be the Maas Mine. The Negaunee Mine now has 400 men at work. A large piece of ground was "let down" at the old Queen Mine and workers were off for two days. It was a good fall for Negaunee's football team which beat Marquette 11 to 0. A rematch was made in late November in bad weather. Play was hard and the score: 0-0.

The Anthony Powder Plant, 1.5 miles SW of town blew up and Sinus Shively and John Nelson were killed. At Christmas, the Savings Bank Store had a decorated wagon with Santa on it and he distributed 1,500 bags of candy to the children.

Sells Brothers' Circus came to both towns in 1884

83

1902

Senseless rumors went around town that the First National Bank had failed. A person tried to take money out and found the doors locked. People showed up on Monday morning and took out $5,000 with no problems as there was plenty of money. They didn't know the man had forgotten it was Saturday afternoon and the bank was closed. Levine Brothers purchased the Big Savings Bank Store.

60. Ten Mine Cave-in Deaths

On Tuesday, January 7th, the Negaunee Mine again caved in 480 feet, and ten men were recorded missing of over 300 men underground. The blacksmith building fell in and disappeared, and the dry may also fall. Jacob Hanala was found dead, and Angelo Ricardo was found alive, four hours after the cave-in. Mine workers and five mules used for pulling mine cars underground were rescued through shaft No. 2. The No. 1 shaft is inoperable with damage in the three top levels. It seems to be a new continuation of the spring of the 1901 cave-in, when an engine house disappeared. The new hole is a "large funnel-shaped abyss" with water and sand freely pouring into it from under the frozen surface which breaks off periodically. Five large pumps have been installed. The other recoveries were reported on 3/14/1902: Louis Mattson, Wm. Williams, John Sullivan, John Pearce, John Pascoe, Angello Cirrelli, Erik Lahti, Wm. Hakkinen, and Julius Anderson. The engine house foundation has been found underground in widely scattered pieces.

At the end of February, the Maas Mine Shaft has a launder to empty the water all the way to the outlet of Teal Lake. It is down 63 feet and no quicksand has been found yet. It is very warm and a professional ice skater cancelled out his show because there is no ice here. Woods workers are also unable to do any work.

The newspaper guesses it is the warmest winter in the last 20 years. Other news was that Louis Sporley sold his double-barrel shotgun with lottery tickets costing only six cents each. Congratulations to Lewis Corbit who got a nice deal. D. G. Stone is selling his business which has 900 customers. He can't walk around so much any more. In Marquette, Peter White has obtained Presque Isle for a city park. North of Marquette a new town is being established and will be called, "Big Bay."

A Kinodrome show is coming for three days and will show a special movie on the McKinley funeral and well-known Americans. Rev. Gilchriese of the Methodist Church went to all the saloons on Saturday night and found several still open past hours. Several men from one bar chased him back to the parsonage, entering the yard and kicking him. Police arrived and six were arrested. The Maas Mine shaft is now down 106 feet but is battling ten feet of quicksand as they go down. The old New York Hematite Mine between the Lucy and the Blue will now be the Breitung Hematite. The CCI is purchasing the Negaunee Mine and six individuals will share the $1,500,000 paid for it. The CCI has also paid $1,500,000 to Mrs. Mass and George Lonstorf for a 50-year lease of property. They will each get $90,000 a year.

The Maas Mine in April was having severe problems with quicksand, which often boiled up the shaft up to 12 feet from the bottom. It all had to be pumped out while men quickly took to the ladders. The pump has a difficult time to handle the sand water, with sand wearing away the propellers. The next week quicksand came up from the bottom at 123 feet all the way to 50 feet from the surface. Five pumps stopped, and land a ways from the shaft has sunk 20 feet.

Dr. Cyr has been confined to his room all winter, but has been out a bit in May. Mr. Barabe has his Negaunee Hennery running with 1,200 hens and gathering 5 to 20 dozen eggs a day. He is selling fertilized eggs and 1500 chicks of several brands of hens for others to build henneries. Sam Mitchell has had his large three-seat carriage overhauled. It will still use horses, but now has rubber tires on it. Louis Kellen is working on the basement for his new building at Park and Mitchell. Mine wages now are from $1.50 to $4.25 a day. Rev. Stanaway is building a seven room home on Snow Street. He is in charge of starting Sunday schools in communities all over the Upper Peninsula for the American Sunday School Union.

One train passing through Negaunee on its way to the Copper Country had three extra cars on it carrying immigrants. The city has received its 38,000 pound steam roller. It will flatten out the crushed rock on streets from the city crusher located at the Teal Lake Bluff. The five Negaunee saloon keepers were each fined $27.00 and told by the judge that he doesn't expect to see them again. The Champion Mine has hired 50 trammers from Austria. Several young ladies had to put their parasols in front of their faces as a bull was entertaining two cows at Case Street and the C&NW tracks.

61. Dr. Hudson and the First Negaunee Hospital.

Dr. Hudson is patterning the first hospital of 42 x 48 feet after the Barabe Home (The Negaunee Historical Museum). The hospital is now the Masonic Temple on Teal Lake Avenue. There was a brief hospital earlier by Drs. Crucial and Swayne, but Crucial left town after embezzling by forgery. Dr. Hudson married the daughter of Civil War veteran, E. C. Anthony.

The CCI shaft of the Maas Mine is making e poor headway in the quicksand of only six inches a day. The funnel caved-in area of the

Negaunee Mine now has achieved a flat bottom and 500 men are at work in the mine. Ringling Bros Circus is coming again. They will be purchasing a lot of food for the troupe and the animals while they are here. There are aquariums this year and sea animals, including a hippo, polar bear, seals, and a sea lion. Negaunee has six high school graduates this year, and Republic has six. With such a good economy, Negaunee's births totaled 223 in 1901. Ben Neely's new store has refrigerators and ice cream freezers. T. Roy has counted in the city, 235 horses, and 462 milch cows, 132 other bovines, 245 sheep and 52 hogs.

Buffalo Bill's Wild West Show will have actual Indians, some who were original participants in battles. After this year, they will tour Europe. The Maas shaft is down 235 feet and thought they hit the rock ledge, but it was only a layer of shale. A band of 20 gypsies came here and collected quite a few dimes and quarters telling fortunes. Negaunee cows made the news when they kept a city visitor awake at his hotel. He found several cows standing under his window all night with bells ringing, but he had nothing to throw at them. He also discovered in the morning that there wasn't even any grass there. A large crowd of 6,500 is expected for the Firemen's Tournament here. 600 cots are available to those who will house some of the town visitors. Houghton and two other towns will bring its city band along. There will be 18 towns represented.

The 8th Annual Firemen's Tournament is all prepared. The Negaunee Race Track has been fenced in at Clark between Kanter and Teal Lake Ave. Games include ladder climbing, hose race, firemen's foot race, flag race, Hub and Hub, coupler's contest, and regulation hook and ladder. The Street RR is providing lots of electric lights "so night will look like day." Michigamme's new fire engine was given a trial run and it was very successful. The Street RR is also providing the rooms at the Breitung Hotel with a two-way bell system with the office. Marquette County paid $439. in wolf bounties last year.

There are a number of typhoid cases and all are requested to boil their water. There were 29 fatal mine accidents last year, with 33 mines having 5,518 employees. The Maas Mine shaft is a bit out of line and only down 139 feet in nine months. Quicksand boils up and it takes 3 days to pump it out. The CCI has taken over the nearby Barasa Mine and is pumping it out also. The Queen Mine caving is now two acres and has taken three homes and some DSS&A track. Chief Kawbawgam died at the age of 103. Many met him at his cabin on Presque Isle up to twenty years ago when a new residence was built off the island. He was tall and straight.

Cave In At Negaunee Mine January 1902

To the left is a photo taken at the January, 1902, cave-in at the Negaunee Mine. The wood at the bottom was on the surface before the cave-in and is for use in the mine to hold up the tunnels which are under tremendous pressure. Photo is from the Negaunee History Museum. Photo below of Street Railway Cars from Art and Anona Quayle and 2002 Negaunee Calendar.

1903

Typhoid can be killed by putting a teaspoon of lemon juice in a half a glass of water according to the Chicago Health Dept. Jack Carkeek, now in the U.S., wrestled George Ringgerford and had a rare loss. The Finnish people here have raised $300.00 to send to Finland to help during the tough times there for rich and poor alike and who are often eating frozen potatoes. The Oddfellows (IOOF) will build a fine building at Tobin and Iron Streets. In the large pit formed a year ago at the Negaunee Mine, a hole has now appeared in the flat bottom and will allow water to drain into the mine and be pumped out.

Bicycles in 1903 will have two speeds available and improved brakes. George Maas and his wife are returning from their six-month trip to Europe. Several motion picture shows are traveling through with shows at churches and town organizations. Twenty Negaunee businessmen have formed a group. John Sullivan, former blacksmith for the Jackson Mine was here visiting, and at 60 he looks well. He said that long ago he "cut out the booze." Most nationalities were seen wearing the green on St. Patrick's Day. In April, Mr. Griffey, the Iron Herald editor, formally retired after 30 years. The Post Office will only rent boxes to adults and youth with parents' permission. Too many young people are corresponding secretly without parent's knowledge. Jacob Salo has a $450.00 electric piano that will play you a tune. Negaunee High School girls are having a game of basketball at the Adelphi Rink. Mahara's Minstrels, a group of black artists numbering 30, are coming here in their own Palace railroad car.

Twenty-five Italians arrived here and went right to work at a railroad construction camp. At the high school, the Negauneesian #2 will be produced by the students this year. Eight will graduate. Mr. and Mrs. Sam Mitchell lost their 17 year-old popular son of diabetes. The city wants to install water meters like Ishpeming. When they found out they would cost $15,000 they said "Let well enough alone." The local firemen have engaged a movie picture company to come and show travel films and comedies.

At the CCI Bellevue Farm on the way to Palmer, 50 steers will now be raised for meat as an experiment. The first wagon death we have read about happened to Mr. Tom Tuuri who was going out to get some wood with his horse and wagon, both of which were also almost totally destroyed. Mr. Kellan's Cash Grocery offers coupons with purchases that can be exchanged for Maple Leaf chinaware. John Schwartz planted 43 bushels of potatoes and will harvest almost 1,500 bushels.

Humboldt will have a large July 4th celebration with a parade through the principal streets and lots of games and music by the Marquette City Band. Many of the Negaunee stores are now installing large plate glass windows with room inside the store for large window displays. And then this notice on June 26, 1903, of the first cars: "A Couple of Automobiles from Marquette Attracted Considerable Attention on Our Streets Sunday."

In July, the Maas Mine finally reached a ledge 163 feet down, only to find that it was on an angle. With two feet of ledge taken out on one side, the shaft sits flat on it. A second shaft will now be started. The problem now is that the launder dumped all of the sand into the outlet of Teal Lake and now water is running into the lake, instead of into the Carp River. The mine notes that only clear water will be flowing into the stream from now on. There is a lot of sand as the level around the shaft dropped 60 feet in the last year while pumping the shaft out.

A recent count of cows at the Maas Mine location counted 173 and "had the appearance of a well stocked cattle ranch." George Haupt has a nice newly painted delivery wagon. John Donaldson one of Republic's heaviest residents died at age 62. His weight was upward of 200 pounds. The Iron Herald, with electricity, shines at night. Skerbeck's Great One Ring Circus will be here with acrobats. It went out of business after leaving here and going to Crystal Falls.

National Mine's Mr. McVichie had a dog that set out to see the world. He arrived quite dirty at the Breitung House, being quite tired as well, and laid down to rest. Later, after cleaning himself up, he was ready to go home. The Methodist Sunday school marched down Iron to Cyr and boarded street cars to Cleveland Park. The procession was a couple of blocks in length and the Negaunee Band provided music at the picnic. E. O. Gillespie, former Supt. of Michigamme Schools, has completed dental college and has opened his practice here in Negaunee.

We note that ore shipments are down and may end before the shipping season is over. However, it is also noted that the mines still seem to be stockpiling at high speed. Captain John Bartle lost his wife and has sold his home here and will leave to visit his children who are widely scattered. The paper notes: "The good wishes of many Negaunee people and acquaintances will follow him." At the MacDonald Opera House, the latest plays have been well attended, such as "The Convict's Daughter," and "The Fatal Wedding." On Case Street, a brand new concrete sidewalk has deep shoe prints in it the entire length. In Champion, the mine has closed and 350 men were released in October. Most single men have left, and the mining

company is allowing those who wish, to stay for the winter, rent free. Ishpeming mines are also slowing up, but 60 men were hired at the Negaunee Mary Charlotte Mine. It has orders for 35,000 tons, but the C &NW cannot find enough ore cars to ship by the end of the boat season.

The newspaper printed the location of all ten fire alarm boxes. A cage at the Hartford Mine fell 325 feet when the hoisting cable broke and had two men in it. Mr. Charles Blumquist lost his life, but surprisingly, John Renstrom survived and is likely to recover. The Mine Inspector's Annual report for the past year shows 23 mine deaths. They include 11 Finns, 4 English, 3 Swedes, 2 Irish, 1 French, 2 Italians. No Americans as such, died.

The city light plant at the Water Works needs to expand. It has sold 1,300 lights, and its capacity is only 1000. Luckily, not all are on at the same time. The city had a choice of joining the Street Railway light plant, but has chosen to stay in business and get a larger boiler at the water plant building. The high school basketball team this year is comprised of two Jackas, J. Toms, C. Calligan, A. Koob, B. Tourville, B. Slimes, J. Trudell, H. Brimeau Jr. N. G. Begle and Sidney Webb. A full-grown skunk is living under the Opera House and is so bold as to walk down Iron Street in the day time. Peter Trudell who had a little store at the old post office has received permission to be at the new one. He was there the next day. Mr. Mucks Meat Market now has an electric "silent" meat cutter.

The fire department has obtained a new wonderful way to put out small fires without all the water damage. It is called "Blaze Killer," and for a trial it quickly put out a fire of lit kerosene burning an abandoned shed. In the next few years, it will be noted that this "Killer" had been used very successfully. At Christmas, Rosen's Store gave the girls pin-trays, and boys, little musical instruments.

Mr. Haupt's New 1903 Delivery Wagon. Neg. Iron Herald

Here is an innocent billboard as seen looking southwest from Iron Street. Please notice, however, the Mine Shaft of the Number 7 pit of the Jackson Mine just off the Street at the top of the photo. The DSS&A Railroad also ran behind this façade. After the mine closed and the pit filled up with water, it was thought of as just a pond, until six-year old Dale LaFreniere drowned in this or one of the other two shafts here on June 6, 1953. Now fenced off to the public. Photo courtesy of the Negaunee Historical Society.

1904

Leap year is this year and the Girls Dancing Class will hold a leap year party and the girls promise that "none of the male guests shall be permitted to become wall flowers." The fire dept. likes it when "fires" are telephoned in as they can tell the size and how best to put it out. Stores are now closing on Tuesday and Thursday evenings at 6:00. Mr. O'Donoghues's new electric Rexall sign in front of his West Iron St. store is attractive. Ten Finlanders have left for Colorado, "to grow up with the country." Cheap Joe's store treated his clerks to a good time straw ride to Ishpeming and back. It is January, but there seems to be little danger of the Cow Bell Club disbanding. The Pittsburg Hockey Team passed through here on the way to play the Portage Lake Team. Mr. Sjoholm's fine bicycle shop seems to have been torched by someone and he lost 20 bicycles, 14 which were customer owned. He has been ill and just getting back on his feet. The Woodmen organization is raffling off a bicycle to help support his family.

In March, a very bad snowstorm occurred and the C&NW got stalled at Union Park. Some got off to walk in the storm, but most stayed on the train. The CM&St.P. which runs from Champion to Marquette on the DSS&A tracks also got stuck in Ishpeming. Schools continued with almost half the students missing. Mr. Sjoholm is rebuilding his bike shop, but making the doorway a lot bigger and says he wants to be ready for autos. At the Opera House, colored exit lights will now mark the exits and the two stairways will be improved. The people at Portage Lake are very happy. Their team won the Hockey Championship of the U. S. in Pittsburg.

62. Dr. Cyr Dies

Dr. L. D. Cyr, pioneer physician here is dead. Born in Canada, he came to Eagle Mills area in the 1850's. He opened up a much-needed drug store, and art gallery, and the first jewelry store. He built the first brick building and stocked it with groceries and candy. He was the Jackson Mine physician and often saw his patients by using snowshoes. He was 74, had a camp at Little Lake, and a home on Cyr Street.

There is the story of one of our young girls who is dating a timid fellow, so she messes up her hair to look like she has been kissed. Mr. Holzhey, now at the Marquette Branch Prison is hoping for a pardon after 14 years. There are ads in the paper for seven railroads with their passenger timetables. Dr. A. C. Smith of the Presbyterian Church has been a supply pastor for three years and has handed in his resignation.

Wm and Arthur Maas have each ordered a new automobile and expect the machines to arrive shortly. Art's will be a very large one, and Wm. will have a small "run-about." New rules for bikes will keep them off sidewalks, and they must have a light on at night, like automobiles. The cows are under control with the city divided into four districts and each has a hired cowboy and loose cows are rounded up It is a late spring and the first ore boat arrived in Marquette the week of May 15th. There are 15 high school graduates this year. The city of Negaunee is working on getting a Carnegie Library. They had a public meeting and all were in favor. Plans were submitted to the east, and the reply was that the city needs to guarantee $2,000 a year, not just $1,000, for the $20,000 building. The city asked the committee to work further on the project. Sam Mitchell has offered a nice corner for it for $2,500. Nothing more is ever heard in the paper.

63. Dr. Hudson Disappears

Dr. Hudson, owner of the Negaunee Hospital, and recently elected Mayor, became lost in the Sands area. His partner, Mr. Miller, has been found. People are going out to search, taking local trains. He was found on June 6th. It is thought he drowned trying to cross the Escanaba River for that is where the body was found. There were four other deaths in June when four people of six, in a capsized boat, drowned in Teal Lake. John Larson and Wm. Buzzo saved two of them.

Paul Honkavara has bought an automobile at his Palmer store to travel back and forth to Negaunee. Druggist O'Donoghue has won a new car from his drug company, and says he has two choices, either to take cash instead, or to rent it to his customers to drive. In the Champion and Humboldt area there was a severe frost on August 6th. Negaunee stores closed from noon to six for the Barnum Circus. The mines closed all day. In 1904, Ishpeming is still larger than Marquette with 11,623 versus 10,665. Negaunee has 6,797. The town council has dispensed with the four cowboys and the Cow Bell Club has reorganized. In Marquette, the new Court House was dedicated.

The County Mine Report lists 15 deaths with a note that this has not been a year of great mining activity in the county. One safety improvement is suggested of miners not handling blasting powder with candles placed "precariously on their heads." The juvenile football team, called the "Invincibles," did not do well in its first game, losing 40 to 0. Even worse, the Herald printed the team members names. The Blue Mine has let go over 300 men. Delangelo Andrea fell down a shaft at the Negaunee Mine over 85 feet and although he has some broken bones, he is recovering fine.

In November, Capt. Sam Mitchell and Mr. Mather visited where the old stump was, marking the first discovery of iron ore. "Mitchell has arranged to have the spot marked for all time by the erection of a monument of masonry and native jasper." The Jackson Forge will also get a large monument. Both will be a pyramid shape. The same month, the Marquette Gas, Light, and Traction Co. bought the Street Railway Company. The George Maas Family is going to winter in Mexico and return when their fine new home on E. Main Street is completed.

It is Christmas time and Kirkwoods has a nice train running around in its window with many small mechanical items also at work from a small stationary steam engine. Lots of little faces are pressed up to the glass. Carl Peterson's store is showing two new modern washing machines. A very large snowstorm hit, the largest since 1899. No trains from Tuesday morning until Wed. evening. The city hired 75 men on Thursday to start cleaning the streets. Some firemen slept at the station in case of a fire. Many were using snowshoes.

The final news is that the Barasa Mine will open on New Year's Day, and that Mrs. Arland will move her millinery shop to the Muck building in the old post office area.

Here is a rare photo of Mr. E. C. Anthony's Low-Moor Farm near Negaunee. Many of the town's men had farms in the summer and lived in town much of the year. The Anthony's daughter was married to Dr. Hudson. Photo from the Iron Herald of 6/4/1904.

1905

Negaunee will begin the year by establishing a Board of Public Works. The city had visited Ishpeming's organization, and such a board works well for them. Ten Finlanders arrived here from Finland. Mr. Griffey, who sold his share of the Iron Herald, is writing about his trip to sunny California in January. It is rather nice here, too. The harbor in Marquette has no ice yet. There will be a nice dance at the Opera House, and the Negaunee High School orchestra will play.

In February we learn that the C&H Copper Mine incline shaft #4 is now down 8,100 feet. There is a "matrimonial bacilli" at the Breitung Hotel as several of the single men have succumbed to marriage. There is hope for three others. The library notes that they now have 26 magazines to read. A play at the Opera House is "Why Girls Leave Home." Several mines are switching to electric motors to pull ore down drifts to the shaft, but the Mary Charlotte lowered four mules down to do the work. A little girl had the accident that happened to many. She got her hand caught in the wringer of a new electric washing machine. She may lose a finger.

64. The Cleveland Cliffs Iron Co. owns the Jackson Mine

March brought good news. The CCI believes that there is much ore left in the Jackson Mine which has not been active for the past nine years. Much diamond drill work is going on. They also purchased a lot of timber land and the entire old Fayette Furnace property. They will also enclose pit #7 with a fence between Iron and the RR tracks. They shipped several thousand tons of ore stored at Pit #7 since the 80's to the furnace in Marquette. In March it is still warm and the city dog races for the kids on

St. Patrick's Day saw poor conditions. There were lots of prizes, including consolation ones to make almost every kid happy. The sidewalks of Iron Street were crowded with spectators.

Mr. Mather of the CCI Company will sell seeds at one cent a package to school children of Negaunee and have a contest of the best gardens. Children ordered 2,688 flower packs and 714 vegetable packs. The Negaunee Businessmen's Assoc. has decided on closing at 8:00 p.m. on Monday, Wednesdays, and Fridays. Saturday nights are optional as are nights when miners receive pay. The Presbyterians chose a new pastor, but he has declined to come here. Almost 2,000,000 brook trout will be planted by the Soo Fish Hatchery this spring.

Lots of people are getting automobiles, and Paul Honkavara of Palmer will be an agency for buckboard automobiles. Dr. Robbins, using the new alternating current city electricity, will install an X-ray machine. The city is now getting quite a few new home customers at the city power plant. The public schools filled the Opera House to overflowing for two performances of the operetta, "Little Bo Peep." In June, graduations from high school only numbered eight, five being girls. As usual, Negaunee Alumni are planning a reception for them. 36 eighth graders will be entering high school next fall. A total of 2,213 children were enrolled in Negaunee schools this past year. There are lots of pianos being sold, with a Henning going to the Odd Fellow's Hall. The Chevaliers d'Lafayette is having a county gathering here with a street parade, picnic, and a ball.

Donald MacDonald died in May, at age 64. He built a mercantile business at Pioneer and Main in 1883, and put the Opera House upstairs. He later enlarged the building. He leaves a wife and ten children. There was a death of a man

who recently visited Canada, from smallpox. Quarantines, vaccinations and fumigations are in effect. A week later a man returning from Lansing had smallpox. A new game law allows a hunter to take only 12 partridges a day and no more than 50 in possession.

Word was received from Capt. Merry, now living in Cleveland. For 20 years he has been diabetic and has had three toes removed. Then he was found to have blood poisoning and the treatment found he was no longer diabetic. (Perhaps he lost some weight.) There have been many mine deaths, and many injuries as things are going full speed at many mines. Peter Johnson died when run over by a night train and had both legs cut off. He managed to pull himself off the tracks into some bushes where he died and was found by some boys the next day. A long series of vibrations were felt here and in the Copper Country in late July, a rare U. P. earthquake. Mr. Maitland is now the Lieut. Governor of Michigan.

The August 4, 1905, newspaper noted that there has been a very severe frost, not seen for a number of years. At the Miller Bros. farm there is a complete loss of 600 to 800 bushels of potatoes. Mrs. Stone has sold her store and is leaving to go with her son. The Kuhlmans have sold their grocery business to H. G. Muck and plan to travel, but stay here to live. John Redness and family left for Tacoma, Washington, two years ago and have returned. Mr. Allison is putting the Adelphi bowling alley in the Sundberg Block. Mr. Cohen has started a new store in the Saladin Building, and called, The People's Store. It held a contest to guess how many people came into the store in the first two days. Mrs. Dell of Palmer guessed 1,500 and won. It was the closest guess to 1,311.

The City has been asked to do something with the entire sewer running into Partridge Creek.

It is affecting the quality of the water in the creek outside of town where the new mines are. Mr. E. C. Anthony and his wife and daughter have returned from Denver where he joined his Civil War comrades in a G. A. R. (Grand Army of the Republic) encampment. September is apple time, and an orchard in Sands has produced 140 bushels of Duchess and Oldenbergs. Chin Hee of the Negaunee Laundry has married Lillie Gorris of Calumet, and the new interracial couple will live here.

The cemetery problem is arising again and the CCI wants the city to accept a 40-acre parcel north of the Maas Mine. It evidently was not accepted. At Halloween, boys were seen throwing peas, but it was girls that were seen soaping windows. There was a very large explosion of the Miner's First National Bank in Ishpeming. Many people walked all the way there to see it as all the streetcars were continuously full. A rare ring-tail cat was caught in a mink trap by Herman Thiele. It was three feet long. Three weeks later an Ironwood man wrote and said ring-tails are usually in Mexico. Mrs. Arland announced that she will retire and is selling her showcases. Big Louis Moilanen, the 8'2" giant from the Copper Country who traveled with the circus, has decided to return home and was on the train coming through town in December. Ten children won garden prizes, with two from each of the following schools: Park, Case, Jackson, Cambria, and Rolling Mill. The People's Store is giving out stamps with purchases, and a full book can be turned in at the First National Bank for $3.50. The Sundberg's Jewelry Store is going out of business. The year started warm and ended warm. Christmas day was very pleasant, noted the paper, and lots of people were out-doors.

1906

People went in debt back then, too. An ad reads: "$3.00 Down Buys a Singer Sewing Machine." The city has a new street plow which works well, but the newspaper noted that the sidewalk plow would work better if pulled by two horses. The Negaunee City Band has had to disband because the members are working three shifts and the group cannot practice together. Tax-paying citizens will be asked to vote in the next election for a new high school. The city has decided to install water meters. With the Temperance movement on, the Finnish Lutheran Church expelled eight saloon keepers; some had been charter members. Gypsies arrived in Negaunee but were not welcomed and left the next day the same way they arrived - by train.

A great calamity occurred, when all of a sudden, the Silver and Iron Street "best water" spring dried up. It is supposed that a new tunnel of the South Jackson Mine tapped into the diamond drill hole artesian well. The winter road has been opened to Eagle Mills, and will be opened to Marquette soon. You will be able to sleigh in February rather than just take the train. You can also go to Fred Merten's Plumbing Shop and see a new tub with a shower. Quite a few people are remodeling and installing bathrooms. Lots of people came to the ski tournament. All the trains had to add on extra cars, the C&NW pulling six extra. You can enter a different kind of ski tournament. It will be a race down the length of Teal Lake and sponsored by the Finnish Gymnastic Club.

Teachers had to be careful in 1906: "Nowadays let a teacher whip a child with even a twig, and there is talk of taking the matter into courts." At the Adelphi Rink, there are roller skating races, with one just for girls on Saturdays. Mr. Tom Pellow has been transferred to Minnesota.

He had just bought the Barabe (Museum) home. Ishpeming has decided to build a YMCA (Young Men's Christian Association), and it is noted that Negaunee should also have such a fine building. The end of the old Concentrating Plant is in view. The CCI has purchased it and hired Mr. Chausee to dismantle it for them. In April, guess what? Yes, Mrs. Arland has re-opened her millinery store now in the Much Block. This was also the time when the great San Francisco earthquake took place. Negaunee people were there, but the T. G. Atkinsons and those wintering were all okay. There was a small fence fire at the cemetery from train smokestack sparks, and as usual, the fire department had it out in no time, again using chemicals.

Michigamme has its own newspaper called, The Thrifty Citizen, mostly for advertisements. On Teal Lake, Louis Peterson is tired of rowing and has a new 1.5 hp boat motor this spring. There were two May days of 80 degrees around the time he tried it out. The Negaunee High baseball team beat the Marquette Normal College team, 14-13. The Finnish Church has installed a bell that weighs 3,300 pounds. May ended with two heavy frosts on the 28th and 29th.

Champion graduated five seniors, and Negaunee, seven. The St. Paul Catholic Church confirmed over 50 boys and girls. A five-foot pine snake was killed in the Jackson Location, after much difficulty. Lots of people are getting new carriages and buggies this summer. We forget that cars only work in the summer for the most part. The paper reminds cars and wagons that they should drive on the right to avoid accidents. There is no law to this effect yet. In from Turin came Mr. James Grimes with a wolf that was 7'4" in full length and 185 pounds without its hide on. In July, the Sons of St. George had a large U.P. celebration and Iron

Street was so lit up at night that the citizens would like to see it that way all summer.

Usually when the mines are all working, the tracks aren't available for circuses. However, in 1906, the following all came to town: The Sun Brothers, The Beatty Bros, The Seibel Bros. Dog and Pony Show with 25 monkeys, and the big Ringling Bros. Circus. In August, the famous rifle shooter Annie Oakley came and gave an exhibition. She once toured with Wild Bill Cody, and could shoot holes through silver dollars thrown in the air.

The court has taken over the Ishpeming and Negaunee, and Marquette Street Railroads, and their power and gas plants and is looking for another buyer. All continue to operate. Mr. Hakenjos, the Silver St. Liveryman, wishes to sell his business and retire. Three railroad carloads of wild horses, however, have arrived and been recently sold here and in Marquette. The purchasers are entertaining the public as they break in their "bucking broncos." At the Marquette Branch prison, there are now 299 convicts being "broke in."

The city managed to get only a little water flowing out of its "fountain," and may get water from another diamond drill hole to the west and pipe it down to Silver Street. The Presbyterian Church is installing an electric motor to run its organ. Dr. Van Riper is now in Champion and many are giving skin from their arms for him to graft on to a little boy who was near death from his clothes catching on fire. The Dr. has hope. The large 100-ton engine, No. 190, is being missed by the public as it is getting an overhaul.

Here is the latest Mine Inspector's Report: 31 mines operating in the County, 5,840 men working. 22 deaths in the past year, 22 diamond drills exploring. He notes, "More activity than we have seen in a number of years." The CCI is moving its farming operation from Bellevue near Palmer, to Chatham where it will be more possible to show what can be done in the U.P. The Street Car line is getting new double-size cars with double-trucks underneath, rather than just one in the center. Something else is new again after being absent for thirty years. Doves are being seen again. Many people are now carrying life insurance. Mrs. Truan just received $2,000. from her husband's policy, following his death.

It is November again, and snow is here. People trying to use sleighs to Ishpeming find that the path made is not wide enough for two wagons to pass side by side and the sides too sheer to move over. They are asking for a two-track road in the winter. A new law has been passed that all can goods sold after January 1, 1907, have their ingredients printed on them on an original label. In town, although summer has ended, baseball in Negaunee has not. It is moving indoors again at the Adelphi, which is said to be one of the finest built buildings in the county. The newspaper had few large store ads or many special Christmas ads. Everyone is busy and probably spending money anyway.

This is the photo of the Negaunee City Orchestra Band. The word orchestra was added when clarinet players joined the band. They are reed instruments, and not wind instruments. There may also have been other non-wind instruments.

Photo is from the Spring, 1977, Harlow's Wooden Man collection. The band was organized in 1913.

We can all thank Mr. Frank Matthews for saving much of the history of mining in our area, from maps he found, to articles of news and history, and mining objects and materials he placed in his own museum on U.S. - 41 until the opening of the Michigan Mining Industry Museum in Negaunee. Many of his items are also at the Negaunee Historical Museum, including this photograph.

LOCATIONS   1925 City Directory

Blue Mine location, south end of Boyer av
Buffalo location, 1-8 mile s of Queen Mine location
Bunker Hill location, see Blue Mine location
Cambria location, south shore line of Teal Lake, 2 miles nw of City Hall
Cornishtown, Ishpeming rd, nr western limits
Furnace location, County rd and C & N W Ry
Iron Cligs location, County rd, ½ mile s of Division
Jackson location, w of Power House
McComber Hill location, cor Division and County rd
McKenzie's addition, ¼ mile e of Queen Mine location
Patch location, s of Division bet L S & I Ry and C & N W Ry
Pioneer location, ns of Iron w of Mann
Power House location, w of Power House
Queen Mine location, Ann, 1 mile e of Boyer av
Regent Mine location, ½ mile s of Buffalo location
Rolling Mill location, County rd nr limits
Teal Lake Hill, w end of Arch
Teal Lake location, s shore of Teal Lake w of Tobin

1907

Logging operations are having difficulty, not because of warm weather, but because there is a lack of railroad cars to carry the logs. Many are being used to carry lumber for the building of all the homes and businesses of the good economy. Home builders are having a difficult time on the other end, getting lumber.

Many old timers are now passing away. Irishman Wm. Conners died at 81. He helped to build the C&NW railroad into Negaunee in 1864 during the Civil War. Another death was the 32 year-old horse "Rocket" of Capt. Foley. When young he was a fine race horse, and had no fear of railroad trains or whistles.. A new ore boat is being launched that needs 32 feet of water, is 58 feet wide, and is 600 feet long. The vote is in on the new high school. Not many votes said the paper, but every one was a "yes."

The Maas and Negaunee mines now have a 600-foot deep safety drift between the two shafts which are a half mile apart. The Breitung continues to stay up-to-date with the installation of modern bathrooms on the second floor. The Adelphi Rink was closed Tuesday night for Lent. The Negaunee boys again held a ski jump at their hill on Teal Lake Bluff and jumped up to 67 feet. 6,000 attended the Ishpeming Ski jump and the longest ride was 110 feet. All the employees of the DSS&A Railroad received an unexpected raise in wages. The Negaunee Bottling works has installed a 2 hp. stationary gasoline engine to run the bottling plant. Mr. Maitland has formed the North Homestake Mining Co. in the Black Hills, just two miles from the famous Homestake Mine.

At the High School, the Glee Club will present the small opera, "Mikado." May Hudson is fixing up her carriage house for both a horse and an automobile. Fred Braastad who closed his large Ishpeming store last year, and was going to leave town, is here and is buying the bankrupt Peterson Clothing Store in Negaunee. One store had this nice ad: "The person who disrupted the congregation last Sunday by continually coughing is requested to buy a bottle of Foley's Honey and Tar." There are also quite a few advertisements of Ishpeming stores in the <u>Iron Herald</u>.

Did you know that there was a planned change in May, 1907, of the Street Car route? When it went from Iron to Cyr it had to cross railroad tracks twice, so plans were to change it from Iron down to Jackson, to Mill to Cyr, like the C&NW. (May 17, 1907)

65. A Movie Theater Opens

Mr. John F. Allison will open a moving picture-illustrated song theater. The Electric Family Theatre will show films from 2-4:30 and from 7-10:30 daily with shows changed twice a week. One film was entitled: "The Miner's Daughter." A second theater opened in July, called the Wonderland will be at the Opera House and will show 2,000 feet of film and it will change every day. Shortly after, it added electric fans and special piano music. The Electric Theatre then decided to change films every other night and have vocal music. The very next week, however, it went out of business saying that they could not get insurance.

On May 26, an inch of snow fell, and a week later the Odd Fellows Block opened, and the high school graduates this year total 32, with 12 having finished the past January. Autos have been purchased by Dr. Sheldon, Alex Maitland (a Packard), and Mae Hudson (a Winton). Lots of citizens are also moving up to nicer homes. There are lots of blueberries this summer. Ed Roy of Cascade has orders from Milwaukee and Chicago for 720, 16 quart baskets, of them. A new law requires all containers for gasoline to be red in color. The word for what we know as a baby buggy is still the English word Go-Cart in the ads. Ads are in the Herald for miners wanted at Silverton, CO, and in Bishop, CA.

The newspaper notes that frequently boys as young as ten are seen smoking cigarettes. Even worse was a mine disaster at the Rolling Mill Mine, when the drum of the hoist let go and the cable rolled off and ten men went to their death as the cage dropped 600 feet to the bottom. The drum worked fine up to then, and worked fine afterwards when the cable was re-installed.

John Shea's store is in debt and has come to an end. However, the closing sale went on forever. It was supposed to be over in a month, in September. In December, their ad read "This will absolutely be our last month in business." It was not. Another theater has begun, called the Bijou, owned by a Chicago man, Mr. Archer and will feature movies and vaudeville. It soon dropped vaudeville for instrumental music. It was in the Sundberg Block on the first floor. Shortly after, in October, a Mr. Allison opened up the Grand Theater on the second floor.

Capt. G. A. Anderson retired from the Negaunee Mine and the employees gave him a fine diamond ring to show their appreciation. In spite of the good economy, 25 Finns passed through on the train on their way to Finland. 12 Italians left for Italy, Idaho, and Arizona.

Peter Trudell, at his little store at the Post Office, is selling the new postcards with a picture on one side, and on the other side, some room for a message on the left, and an address on the right. There is night school in Negaunee and by the second night, 98 had enrolled, with the average age of 26. By November enrollment was averaging about 130. Because of shift work, not all can attend every class,

every week. The two teachers are Mr. Squire and Mr. Erickson.

There are financial problems in the country, and banks and employers are beginning to issue checks in Negaunee. Soon there were lots of counterfeit ones in "large" circulation. Duluth found $30,000 of them. Lots of people are also getting pianos. The high schoolers are planning a yearbook, called the Negauneesian. The last one was in 1903. And who is closing up again? Well, it's Mrs. Arland. She says she is going south for the winter.

It's a warm Christmas and mild temperatures above freezing. There was only one small picture of Santa Claus in a bank ad.

Here is the Iron Herald Printing Press that newspapers were printed on. The paper was placed on the revolving cylinder and the two or four pages of type lay flat on a form underneath. As only the part of the cylinder touching the type required high pressure, it was much easier to print a page than pressing down on the entire chase of type at one time. The cylinder did not move. It only revolved and the chase of type moved under it.

The Iron Herald Print Shop, after 1910 was in the basement of the old Negaunee National Bank Building where it still is located. (U. P. Digitization Center)

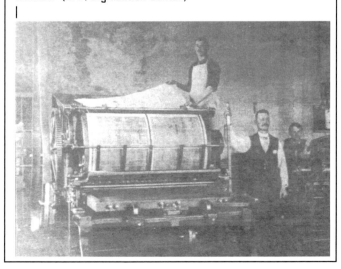

Mr. Rytkonen and a partner built this fine theater at the corner of Pioneer and Iron Streets in 1911. It was called the Star and offered the first sound films and many of the big names that are still remembered, such as Mary Pickford. When he built his new Vista Theater, this was remodeled into an auto showroom. Photo from Marquette County Historical Society Files.

Here is a Mine drawing of the Number 7 pit. I have sketched in where the present Heritage Trail leaves the Senior Citizen's area and goes around a fenced in area, including the Jackson Bowl. You can see the great underground workings in that area as well as south of the pit. The most interesting item is a tunnel going towards, or under, Mr. Merry's home. The drawing of the home is approximate. It has long been argued that he used the tunnel to go back and forth to this pit and its shafts. It will also be argued long into the future. I found nothing in the Iron Heralds. I want to thank Mr. James Thomas for this map to share with our readers.

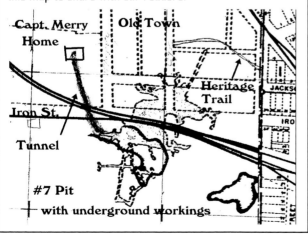

1908

Mild temperatures continued into January and wagons were still in use on the streets. On January 17, the paper reported that it is seldom zero degrees and that navigation is still open on Lake Superior and on the St. Mary's River. Hunters and loggers report that they are seeing quite a few bears as they have not hibernated. In late February the ski races were held at Teal Lake and two spectators fell through the ice into five feet of water near the shore.

The city of Negaunee has hired an electrical inspector to see that all wiring is according to code. "O'Donoghue's Drug Store now has moving pictures," said the ad. A closer reading shows that it is a new electric display stand going around and showing the postcards they have for sale. "Fear not, Bijou!" The Bijou would like to find more vaudeville acts, but they are hard to find. Movies only, now. The LS&I is building its new railroad track from the Teal Lake Hill area down to the Jackson Cut area to join their southside track. There will be two large spans, and some homes have to be condemned in the Jackson Location. The CCI and the C&NW are no longer selling cut-over land, but planning for the future by replanting.

The U. S. Attorney has stated that the Finnish people are Mongolians and not eligible for citizenship. A judge in Duluth has declared they are white and eligible. The Michigan Gold Mine is busy working day and night. The Wonderland Theater will reopen, but has to close on nights that the Opera House has other special shows and balls. John Shea's Store is still having its "Final Clearing Sale." An Interurban Street RR between Negaunee and Marquette and right-of-way has been laid out. Work was to begin in the spring. The dog race in Negaunee is so popular that "wintering" citizens read about the town in newspapers of Los Angeles and Buffalo. Lots of bowling scores and columns are on page five each week from Republic and Michigamme. There is one from Champion, also, on February 28. By March 6, Shea's is closed and Mr. Axel Rasmussen has a new store there with special gifts for the ladies and an orchestra for the Grand Opening.

The water meter report is that people are using less water and the financial collection pays the electricity as the pumps do not have to pump as much water. The Negaunee Greenhouse between Gold and Silver will enlarge. The girls at the Negaunee High School have formed two teams. Mr. Wm. H. Israel, a photography graduate, has purchased Bergland's Studio. Carl Waldemar, the first watch and jewelry store owner in Republic, has died. The government is sending a Washington, D. C., person to the U.P. to begin a campaign to exterminate wolves, estimated to be at 200.

Stores are constantly moving. Here is an April 17 scenario: "The Rasmussen Store has moved from Kuhlman to the Shea Building. Marietti has moved to the Kuhlman Building from the Hogan Building. Boyer has leased the Hogan Building." There are 500,000 non-English speaking people in the state currently and the new Communicable Disease Pamphlets will be printed in several languages. On Good Friday, the Negaunee Band appeared on the streets and played. Mr. Trudell has retired from his store at the post office, and Mr. Phil J. Hogan will continue it. Mr. Sjoholm, the bicycle man, is now also selling typewriters.

The Iron Market is slowing down. The Republic Mine wanted to go to a four-day week, but the Company wants miners laid off, singles first. The Empire Mine in Palmer, all fitted up to start in the spring, may not open. In May, there are still no large sales of ore.

66. Deaths of Sam Mitchell and Peter White.

In May, Negaunee's Sam Mitchell came down with pneumonia at a Chicago Hospital and passed away. He was 62. His will left the family home to his wife and her right to ask for a third of all the estate, rather than take the $6,000 given outright. A month later, Peter White passed away on June 6, while in Detroit. He was born in 1829 and lived here before any trains served the area. He had just paid for installing a statue of Fr. Marquette at Mackinaw Island.

Marquette and Ishpeming are both backing a fine Fourth of July event in Negaunee. Tom Kirkwood will have a new agency for selling autos. The factory is in Kenosha, WI, and the car is the "Rambler." The Copper Country is trying to have the third largest city in Michigan. To do it, however, they have to combine Calumet, Mohawk, Red Jacket, and Laurium and some other towns. It is now June and no ore boats have come yet. Ore sales are only half of last year. A cow tried to get into a yard with luscious green grass in it but didn't do so well. The fence didn't do so well, either. The fourth of July did very well, however, with a large parade of floats, and bands, including Ishpeming's. There were decorated carriages of officials, several nationality organizations marching, and lots of decorated automobiles. Businesses planned a large electrical display, and fireworks at night on the McComber Hill, south of the depot. Some floats were pulled by two teams of horses, and Mr. Braastad's was pulled by an auto underneath it.

Mr. Sjoholm has sold his auto, bicycle, and repair business, but kept the building, and will retire a bit. At the Union Depot, the agent could not get the safe open. They brought in an inmate of the Marquette prison and after a few attempts it was open. The Breitung Hotel and schools will now be equipped with fire escapes.

67. A New Cemetery

Ore being under the present cemetery, and it being on leased CCI land, the city decided to leave it and is looking at a tract near the Carp River. Both cemeteries will be moved. The bad points are: the street car line may not be extended, and the distance is quite far, and funerals will cost more. The road from Dickinson County is going to be extended to Sawyer Lake and Marquette County. We will have to join it from Republic south. Paul Honkavaara of Palmer has a new Ford dealership.

The Annual Mine Report shows 40 mines operating, 5,362 men employed, with only 16 deaths last year, a new low. Fencing is being put around abandoned pits. In October there was a bad head-on accident of passenger train #1 and an ore train a mile west of Greenwood with one death and many injuries. There have been many other deaths and serious injuries of railroad employees of smashed hands and falling between cars and being run over several times. The new Republic High School is completed with steam heat and electricity. A number of ladies dressed up as witches with brooms included and visited their friends on Halloween. All had a good time.

It was a bad year for forest fires and many hunting camps are being rebuilt. Many animals were also driven into towns, especially bears as there was little food. Most were quite startled to see civilization and city streets. A sure sign of fall was when the Sunday passenger trains were discontinued, and bowling and roller skating

were underway. In November, the steel bridge of the LS&I was completed over Teal Lake Avenue.

A debate has ensued of where the new Jackson Monument should be. S. L. Barnes, an old citizen, says it should be further east. Mr. Tom Pellow says the place for it are okay. Football is as popular as ever. There is now a Juvenile Team, and two other teams called the Cubs, and the Teddy Bears. The Republic Mine has obtained newer, improved, acetylene lamps. John Allison sold his Sundberg Block and a building across the street, and the LS&I are putting up "the large steel structure" over the Street RR, the C&NW tracks and Bluff St. Farmers are busy plowing fields for spring in early December.

It is more "Christmassy" this year with Bazaars, church services, and a ticket for the Bijou may win you a $25.00 gold watch. There are special Christmas train excursions, and the greenhouse is advertising Christmas flowers. People are getting pianos.

Here is a share of stock of the Barasa Iron Mining Company. They raised a million dollars by selling 40,000 shares at $25.00 each. This stock certificate was never used and is at the Negaunee Historical Museum.

1909

Weather-wise, the year must have started out as an average one as nothing is mentioned for some time. What is mentioned is illness. Schools did not open as there were many cases of diphtheria and measles. The library was also closed. The town of Princeton is busy growing with 24 new homes completed, and a volunteer fire-department organized of 18 men, to work with Gwinn. They have a Home Comedy Club as well and presented a "farce" at the New Riverside Hall. Last year in Negaunee there were 118 deaths and 261 births. The theater at the Opera House is now the Family Theater.

The city has had problems with its electric plant, and Westinghouse came and remedied the poor service. It has to work well for the lights at the new High School. The main street in Sidnaw burned to the ground and people stood around and watched. They had no fire department. Some individuals many years ago invested in the Escanaba River and Land Co., and it turns out that much of their land has iron ore under it which the CCI is leasing. A total of over $100,000 has been paid out in the past four years.

In February we do learn that the weather is still warm and that sleighing is about impossible. Henry Pascoe never saw one so mild and said the Jackson Cut near the Mine office used to have so much snow that you could only see the stacks of the engines. Several people went to the Chicago Auto show and ordered new cars. Two miners were entombed at the Cambria for 28 hours until reached by a new drift. It is 1909, and there is still no road to or from the Copper Country to L'Anse. They are writing to the papers here to petition for one. They have cars but can't leave. For Lent, the St. Paul's Episcopal Church will have their services on Fridays at both 4:00 and 7:30 p.m. The Chicago and North Western is replacing the old hand-pumped section workers' cars with ones with gasoline engines. Two crews now can do the work of three each day.

Home building in Negaunee continues in spite of mines not running fully. The Cyr tract sold 17 lots, with the old Cyr home having a larger one. J. Rickard, C. and H. Kronberg, A. Carlson, and J. Freschette are building. The city has also accepted the Collins Tract. In March, at the Teal Lake Bluff an 8 year old, Albert Yackel, jumped 16 feet and 72 youth participated under the age of 18. The dog races went well and three new prizes went for the homeliest dog, the smallest dog, and the slowest dog. Raffles are still popular, and Bert Balcom, the chimney sweep, won a team of driving horses. The M. E. Church put a large basement under its old church, 9'6" high and a fine hardwood floor. They may also add a toilet room. There were two pages of printed names of people who were given food and wood this winter. People got Easter stamps in the mail this year from the Anti-Tuberculosis League.

The city sells electricity. They are also now selling vacuum cleaners for the ladies, at the City Hall. Many were purchased. There are eight mild cases of smallpox. The library reports that there are now 1,502 patrons and that all books are being collected, and those from infected homes will be destroyed. Kirkwood's is selling fancy goldfish for fifty cents each. Mrs. Godin has sold the Bijou but must have kept the projector as she is showing films to small towns. The Negaunee firemen sent the steamer by railroad and went with it to help put out a large fire at New Dalton, south of Marquette. The Street Railway gas plant offered to install gas stoves free upon purchase if they lived close to the gas line and the Street Railway. Many took advantage of it. There is no "220" electricity yet for large electric stoves, but many are buying and running "hot plates."

68. The New High School

In June, the new Negaunee High School was dedicated, with a special 8-page section in the newspaper. There are four floors including the basement. Lots of photos, and a fine new gym and auditorium were built. Mr. Mather, head of the CCI, came all the way from Cleveland to speak.

The Street Railway is leasing Union Park, and they are also putting three steel rowboats on the body of water at Cleveland Park. The 1909 Negauneesian is now for sale. The Bijou is showing a film of the Johnson-Burns Boxing Match. Large crowds attended the fight movie on both nights. Those wishing now to become citizens must verbally answer all the questions without an interpreter. Dr. Haidle has installed a ten-gallon hot water heater. Barber Honka has added an electric fan in his barber shop.

The city has selected for the new cemetery, the 80 acres owned by the D. MacDonald Estate and an additional small parcel to the west. It will be by the main highway to Eagle Mills and will be eight times larger than the Main Street one. The city also plans to build a new brick fire hall with a clock in the tower. It will be in the Pioneer Ave. area, south of Iron St.

34 of 40 Collins Addition lots have been purchased. There are new state speed laws for autos, 8 mph in town, 15 in residential, and 25 in the country. The DSS&A will have a new route into Negaunee leaving at Eagle Mills and following the LS&I route, and rejoining their tracks near the old Pioneer Furnace.

Another letter in the paper from Houghton County is asking if a road could at least be built to Iron County so they could get here, saying they currently are "prisoners in a restricted territory." The Negaunee "Fearnot" Ladies will entertain and play baseball with the Chicago "Bloomers." Do you want to know who won? Negaunee won, 8 to 7. The mines have really picked up and are hiring. 200 men for an added shift at the Negaunee Mine will be hired, and 150 more will be hired at the Breitung and Mary Charlotte combined. Steam shovels are busy removing former unsold ore from stockpiles. The Chocolay Furnace, closed for 24 years, will go into "blast." Gwinn now has a column in the paper in the August 27, 1909, issue. New banks here are the State Bank of Negaunee, and the Negaunee State Bank.

Mr. Nicholas Laughlin died. The church was "filled with sorrowing friends," and it was stated in the obituary that "he helped many customers through periods of adversity." As baseball ended in September, the Negaunee team beat Ishpeming by one game. It was a torrid season. Marquette also threatened at times.

Back to Teal Lake water: the state says that all water used in making coffee, drinking, or baking or cooking, should be boiled before using. As October came, the school system had to open up one more first grade room at the Park School, in order to limit all classes to no more than 40 students. And it was time for the Inspector of Mines to give his annual report. 23 men killed. (11 Finns, 4 English, 4 Italians, 1 Swede, 1 Danish, 1 French, and 1 German.) Deaths at 13 mines. 11 of the men were from Negaunee.

Mr. A. P. Swineford died. He once lived in our county and was in charge of the Mining Journal. He left here and became the first territorial governor of Alaska. The Bijou now has a projection room that is fire-safe, and announces that all films now being shown are nonflammable. Hunting season has arrived and the train to town was two hours late, having stopped continuously to let hunters off in the woods. Up to December 4th, farmers have been plowing their fields.

69. More Negaunee Caving

F. Condello, O. Mattila, P. Mundi, and V. Makki were trapped at the Negaunee Mine on Monday, December 13, when they set off a blast and it opened a huge rush of mud into a sub-level, 570 feet below ground, which continued for some time and filled the main drift to within 500 feet of the main shaft. The area was just 800 feet from where the ten men died in 1902. Although three were found dead the next week, Mr. Condello was still alive, standing upright in a space that had continued to get smaller. He had survived for seven days eating bark from a mine timber and drinking dripping water. He walked from the shaft when arriving on surface.

For the children, Santa will have his headquarters at Andrew Ericson's Store. Bring your children to meet him.

To Cave the Surface. *10/4/1913*

The Cleveland-Cliffs company has fenced off the old highway leading past the old Negaunee cemetery, and will soon commence caving the surface of this locality for which it made preparation by constructing a new highway several months ago. The cost of removing the old cemetery was a big one, probably footing up to a quarter of a million when everything has been figured. This is not taken into account by the tax commission. The working of these ore deposits will be a fine thing for the people of the city of Negaunee who depend solely upon the mines for their livelihood. In this country we have to follow the ore under towns, cemeteries, lakes or wherever it makes. It is the ore we are after and we must have it in order to do business.

1910

Condemnation proceedings will continue against Mrs. Donald MacDonald regarding estate lands needed for the new cemetery. She had tried to stop them, but the Michigan Supreme Court voted to deny her motion. Mrs. Arland is still in the news, too. She opened a new millinery store in September of 1909, and she is closing it again after Christmas. More news at the Negaunee Mine when the dry burned down and 300 men lost their underground clothing. They were given $5.00 each to replace the same.

Pianos are still a big item and the Cable Piano Co. has a store now. The pianos cost about $175.00. Wm. Palmear of Humboldt died and it was noted in the paper that, although blinded by a mine blast 40 years ago, "what he was able to do in business and politics during his life was marvelous." He was known by all as "Uncle Billy." There was no mail for the first time in many years when a large snowstorm hit. The C&NW just turned around at Green Bay and went back to Chicago. Swan Carlson committed suicide at age 29. He wrote a note saying that he deemed his life to be a failure and that he thought it best if he took himself out of the way. It was too bad that he didn't know about forgiveness.

Ski jumping is still a professional sport and the large tournament this year will award $250.00 to winners, as well as prizes for the amateurs. Mr. Braastad has bought Mr. Laughlin's inventory and will run the business and provide delivery. A Native American in the L'Anse area recently died, and had the name "Nigani." 300 kids were at the recent Bijou Theater Saturday matinee. A few weeks later, there were four Saturday shows, and even then, people were turned away. Mines are still successful with the Lucy now being unwatered and will be hiring a full double shift. The library is having some problems, and with the exception of Ladies' Home Journal, and Good Housekeeping, no other magazines will be allowed to go out of the library. People are removing parts of some pages. The City Band held a concert, but not too many people came. February doesn't seem to be as popular as the summertime. An ad by J. M. Perkins advertises a large selection of phonograph records and all styles of Edison phonographs.

The first white child born in Negaunee, Mrs. Mary Reidy, died on February 13. She was born in 1853, before the city was even platted out, and only nine years after the discovery of iron ore, as Mary Laughlin. The Carl W. Cook Stock Co. came to do vaudeville at the Opera House all week in February, and took time out to go and see the big ski tournament on a large sleigh they rented. Up to February 17, the weather was mild for a full two months. On that date the readings for the next week were -19, -28, -18, -8. -6, -28, -22, and -20.

The new cemetery is moving along and the CCI has completed the site by removing dozens of trees and stumps, and tendered the site to the city. It is a total of 100 acres. The present road cuts across the S.E. corner of the property, and it is 2.5 miles out from town. The present cemetery is so full that only single graves are now available. Mrs. C. D. Dyer sued local saloonkeepers for $10,000 in damages for the liquor given to her husband that killed him. They have settled with her and each will pay her $125.00. In sports, when the basketball season ended in March, Negaunee was the local champion. However, the next game was with Calumet and they lost 60 to 21.

It has been a long time since we talked about the old Union station on Gold Street. Well, it is still there. The newspaper gives one more push,

saying, "Ramshackle Union Station must go." They noted that the holes in the wood platform have been simply covered with pieces of wood. The health department complained to the railroads about the bad sanitary bathroom conditions. The answer has been to lock the doors of the same. Well what else do we see in March? It is Mrs. Arland opening a store again.

70. Free Mail Delivery

The city is preparing the number system for all buildings in preparation for free mail delivery. At first, it was decided to assign a street number for every 20 feet of curb. It was soon learned, however, that there is a good system of allowing 100 numbers per block with blocks then numbered 100, 200, 300, etc. Numbers on buildings are to be three inches high and street signs that will be installed in town will be the colors of~~~~white on blue.

Alex Bean, filling in as a teamster, was kicked by the horse and then trampled under by it and his recovery is doubtful. In Humboldt, ten boarders at the Hendrickson House became ill. Dr. Van Riper was near by and discovered it was not the food they ate, but the water they drank. All are recovering. It is now the end of March, and there was a fine day of 60 degrees In the spring election, Marquette County voted to stay wet. Some counties are going dry, but the counties next door get wetter.

The Negaunee National Bank displayed the plans for its new bank in the paper. There will be mahogany on the main floor, and two bathrooms on each floor. The bank is paying 3% on Savings Deposits. Counterfeit $5.00 bills are in circulation, all with same serial number which was printed in the paper. The CCI is busy fixing up the Fayette homes as cottages in their new resort there. In the sky at 4:30 a.m., if you get up, or stay up, you can see the tail of Halley's Comet. Remodeling of the Bijou will raise the floor at an angle so that people can see around the hats being worn. The office will also be put out on the street area, allowing more room for seating. Sam Barney, the man concerned about the location of the Jackson Stump, has died. He was 80 and was a worker at the Jackson Forge, and an oxen driver of wagons of ore going to Marquette on the Wagon Road. He was buried in Marquette.

The Scandinavians are planning a large Swedish Carnival on the weekend of the longest days of the year, called: "Varnamo Marknad." A new book is out about "The Mouth and the Teeth." At the Opera House, Barnum, the hypnotist, excelled and capacity audiences were there in spite of bad weather. The Negaunee Rod and Gun Club is planting 40,000 fry trout in area streams. In May, the DSS&A was unhappy with the CM&St.P using their tracks to the Copper Country and to Marquette from Champion, and have instead invited the C&NW to use their tracks in both directions from Ishpeming and Negaunee.

Up to now the city of Negaunee has had two fire companies in two different areas. With the new fire hall, there will be only one, and so they are merging. The new organization held a ball and has already raised $125. The John Mattson farm, south of the city, is being explored for iron ore. The ground is dry from a winter with less snow, and there have been several dangerous fires again this spring. There is another inventor in our midst when Mr. George McEchron invented a flour sifter (explain this to your grandchildren) that works with one hand by squeezing the handle. A firm, he says, is already making them, and they will be for sale in Negaunee shortly. Districts that do not have a high school must pay $20 for tuition to schools to which their pupils attend.

A large fire burned down three business blocks in Ishpeming on Division Street. The steamer and horse team from Negaunee made it in 16 minutes to help put it out, and several men got there from here using automobiles. Citizens were able to view the plans for the new Union Station. It will be 22 x 112 feet, one-story brick with two toilet rooms, with the men's used for smoking as well, and the ladies' can be also used as a waiting room. There will be a large concrete platform to the west of the station.

71. A Downtown Clock

The new clock for the firehall tower is here, and will show on all four sides of the tower. The two present bells will be melted down for one new large copper bell. A Chicago expert is installing it. The public says it is smaller than expected. The dials are 3'8". Also new in town are two black men who are doing a fine shoe-shining business. Another new thing in town for June is the fact that snow fell on the second of the month. Not enough to make the ground white, but very unpleasant.

The Street RR is sprucing up Cleveland Park by getting new boats for the pond, more seats for the many picnickers, and new swings for the young and "young in heart." Negaunee now has three fine brass bands, the newest being an Italian one. Game wardens were awfully busy, and obtained seven game law convictions, and 107 fish law convictions during May alone. 44 seizures were made, and $1,424.00 was paid in fines.

July 4th is coming and there will be some new races, including a fat man's race. You have to be at least 200 pounds to run. A greasy pig is going to be released and anyone who can catch it, going up Iron Street from Cyr, can have him. The next week, after the 4th, we see that so many people showed up in front of the pig box, that he couldn't even get out to run up the street. We wonder if mining is slowing down again when we find the Ohio Mine near Michigamme is closed until there "is more demand for ore." Charles Johnson, the Pabst dealer, is getting his beer in barrels, and bottling it himself. The public is asked to shut off their water sprinklers when the fire alarm goes off so that the firemen will have plenty of hose pressure.

The town is enjoying commercially made butter by the Keystone Creamery of Champion that has installed an electric butter making machine. The janitor of the Catholic Church, when it was at Brown and Main, John Stecher, died. At age 18 he had lost his right arm, but learned to play the cornet for several bands, and made a living by also being a custodian at the Park School. Two men drove their automobile from Cleveland, Ohio, to Escanaba, but took a train from there to here.

72. Iron Street gets Paving?

Iron Street will get concrete curbing, gutter, and catch basins will be installed. Plans are then to have a layer of larger rocks, then five inches of concrete, then sand, covered with creosoted wood blocks. It's not what we expected, and I am not sure it ever took place. We shall soon read that tarred roads will be coming into being in large numbers.

For the younger readers we should tell them about August Raatikainen's fine horse, 'Jerry,' the large gray horse that has died, When August got to town with his pop wagon, he would get off and after carrying in every load to every establishment, the horse would move on to the next stop, be it across the street, down an alley, or whatever, all through town. After the last stop he waited for August to remount the wagon. "All will be sorry to learn that he will be seen no more." In late July, Negaunee once

again had some typhoid cases, and Teal Lake water will again be tested. A wire fence is being put all around the lake as well, being 5.5 feet high. Campbell Bros. shows are here and arrived on 42 double RR cars. There was a street parade at ten a.m., and one performer did bicycle somersault across a gap. At the north end of Tobin, a Socialist Labor Temple is going up that will seat 500. At the farms, hay is poor this year, and one farmer sold his for $24.00 a ton and the buyer will cut it and haul it. D. T. Morgan drove his auto all the way to the county line, and was the first car ever seen at Witch Lake. They are still building the bridge over the Michigamme River which is held up with a large cement pier in the middle of the stream. A week later, a group of automobilists made a round trip to Iron Mountain, using the new bridge.

Mr. John Mattson had two wolves at Eagle Mills come into the area of his farm buildings so he shot both of them for a bounty. He hears many wolves in the woods. In August, the city water tests came back, showing the spring water at Silver and Iron is about the best there is, while Teal Lake is okay, but has organic matter that could grow a disease. People will again have to boil all drinking and cooking water. St. Paul's Catholic School has paid off all its debt by raising $2,740.00 with its bazaar. Hanson and Son will install a large electric refrigerator for their meats, for the public, and for other dealers to rent. It will even make large blocks of ice. Improvements are happening at the Bijou Theater as well. They now have two picture machines so they can show reels continuously with 3000 feet per show.

Baseball is big in Negaunee with many teams. The professional team is the most popular. This year it won the area with 24 wins. Marquette had 22, and Ishpeming and Crystal Falls had 12 each. To get the championship, they had to play Escanaba, with three games in each town, and won in both. A large crowd of about 1000 met them at the train station at 11:00 p.m. The owners of automobiles in Ishpeming and Negaunee are forming a club. George Maas and friends have formed the North Range Mining Co.

73. Graves Being Moved.

The Stensrud Bros. were hired in October to begin moving graves from the old to the new cemetery. Price of lots will be $10 to $40, with people being invited to visit and choose their new grave sites. On the appointed day, some were there are 4:00 a.m. and several hundred by 9:00. Marked stakes were driven in as lots were chosen and deeds were rapidly given. Sites along main driveways will cost more, and there was no rush to get plats by the county road. Roman Catholics were invited as well.

Bodies will be carefully removed, and headstones will also be moved and repaired if in damaged condition. Bodies with deteriorated caskets will be lifted carefully with a board placed underneath. Rakes will be used instead of shovels. All bodies will be transported by auto trucks. The new lots will be larger than the old ones. A building is being built at the present cemetery in which carpenters will build the new wooden rough boxes for all bodies. It is planned to move 50 bodies per day. Records are being kept of the condition of each body.

Quite a few people are removing their graves from Negaunee. The first month, 50 returned bodies to Ishpeming, as originally they had no cemetery there. The Hoch family will all be moved, along with a grand 7-ton tombstone, to Milwaukee. Their son died in the Pendill Mine Shaft on October 14, 1892.

The first burial in the new cemetery occurred in late October, when Mrs. Catherine Brand died. She was living when the original Catholic Cemetery was at Main and Brown (The Museum lot). By the end of November, the Stensruds were working on the two old, unknown, portions of both cemeteries. Many will go into nameless graves, but every small fragment is attempted to be saved. A nameplate was found on a coffin plate, Edward Lampress, 45. No coffin and only a few bones.

There were 29 mine deaths this year, mostly by a fall of ground underground. 48 mines operated in the county this year and 6,546 men were employed. The mine inspector ended his report with a note that shipments went down at the end of the season. We are coming to the end of the story of the great Negaunee World Wrestler, Jack Carkeek. He has been arrested in California with a false name and arrested for setting up fixed wrestling matches for money. It was found to have been going on for many years, including the English matches. When living here as a boy and young man, he was a "crack" baseball player for the Negaunee amateur baseball team. He will be on trial in Omaha.

Free mail delivery is planned for February. There will be an exam to fill three carrier positions. They must be 18 to 45, except if honorably discharged from the military. All who wish to gain delivery must have a locked outside box. Mail will be delivered four times a day, at 7:10 a.m., 10:40, 2:40 and 5:00.

In December, it was also announced that there will be a hospital, built five miles east of Negaunee, to treat tuberculosis. It will be at the old town of Morgan, with electricity supplied by Marquette.

In the middle of December, 3,370 bodies have been removed. There are some heavier snows, but work continues. Two-thirds of the bodies removed have been from the Catholic Cemetery, with possibly 1000 left there yet. Mrs. Sarah Amelea Cassidy Fay was removed easily even though the casket had disappeared because her body was partly petrified. She had suffered greatly from rheumatism and died in 1895 at age 45. On the Morisette family lot, a bottle with a child's name, age and date of death was found.

It is Christmas, with much celebration. There was a Christmas Festival, and New Year's Eve Festival, a New Year's Eve Ball, and the yearly German Ball. The newspaper used a full page for a scene from Dickens' Christmas Carol and which said, "God Bless Us Every One." The new Union Station opened up with a RR ticket office on both ends and a large sign that said, "To the People of Negaunee, Merry Christmas."

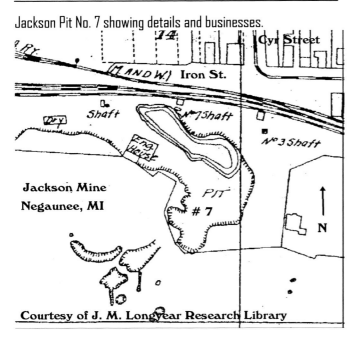

Jackson Pit No. 7 showing details and businesses.

Jackson Mine
Negaunee, MI

Courtesy of J. M. Longyear Research Library

One of the early children's automobiles. Just like Pa's. Photo from the Negaunee History Museum collection.

1911

The U.S. Post Office didn't know about the U.P. winters. The free delivery of mail will not start in February, but begin in May. The three letter carriers have been chosen and are David Murphy, Louis Houle and Oscar Holmberg. After delivery began, the routes had to be adjusted for heavy downtown loads. Sunday service at the post office was abandoned, but the boxes still in use will be available. It was only a matter of a month or so, and the newsstand there was for sale. "Bart Black" Holzey (Holzhay) has been released from the Marquette State Prison after 21 years and being captured in Republic. He is said to be a changed man after an operation. Lumbermen in the woods are having a difficult time again this year. Now it is because of the deep snow, three feet of it. John Honka, the barber, will occupy the basement of the new Negaunee National Bank which is now in its new building. With the new city hall being built, it is recommended to move the library from the second floor of the present city hall to the old second hose house of the Fire Department. There was tragedy in Ishpeming to start off the year. A large explosion at the Pluto Plant in National Mine killed ten men. Body pieces were picked up for two days.

Mr. John W. Goudge of Negaunee continues to practice his hypnotism, and this week, for the first time, gave a show here as "Professor Wicks." He was very successful and "made good!" D. T. Morgan, the State Representative from Ishpeming, will introduce a teacher retirement bill that will give a half of the last five years average salary as a pension if they have taught for at least 30 years. It must have passed. At the new Labor Temple, 500 miners and citizens met to discuss an 8-hour work day.

Wm. Cote and some friends jumped off a mail car as it entered Marquette as they were riding illegally. Cote caught his foot and swung under the car. His remains were badly mangled. The other bad news is that not much ore has been sold yet by the mining companies for shipping this summer. In town it is spring, and businesses are moving once again from building to building. (Moving on up!) We haven't mentioned food prices for a while. 50 lbs of flour is $1.50, butter is 30 cents, a dozen eggs are 20 cents, 25 lbs. of sugar is $1.40, and ten bars of Fels-Naptha soap is 45 cents.

A new Lake Superior and Southeastern Railroad has plans to purchase the right-of-way of the old Iron Range and Huron Bay RR, and take the timber from along the route. That is evidently the beginning and the end of this story. Negaunee is planning another professional baseball team of four towns this summer. The fourth team this year will be Escanaba.

We all know that Mr. Rytkanen was the builder of the Vista Theater. In 1911, he is building his first theater at the corner of Iron and Pioneer. It will seat 800 and will be the largest in the U.P. With nice weather in April, boys are out catching trout. The game warden found four of them with 100 fish. They each paid $6.00 and the fish went to the Morgan Heights Sanitarium. In Republic, there are now four automobiles. The Stensrud Brothers are still busy again at the cemeteries, and are moving up to 75 bodies a day.

74. Six Die in Hartford Fire

A big fire occurred in the main shaft of the Hartford Mine near Teal Lake and six men died. Five were at the Hartford: H. Dower, E. Puska, H. Wherry, R. Eland Sr., and A. Frederickson. John Tamblyn, working at the Cambria Mine next

door, died from poisonous gas that came through a connecting drift. Several others were taken to the hospital. The 1910 census figures are in, and Ishpeming is still larger than Marquette, 12,448 to 11,503. Negaunee is third with 8,460. The Republic and Forsyth townships have over 2000 people each, and Champion and Tilden Townships have over 1000 each.

By June, the removal of bodies at the cemetery was near an end as state law forbids the work in the heat of the summer. To double-check for bodies, the cemetery will be trenched from one end to the other. The body of Mrs. John Truan, buried just 15 years ago, was also found petrified. On Memorial Day, many who went to the cemetery concluded it was not as far out as they thought. The DSS&A brought 240 people out and stopped as close to the new cemetery as it could, with a path built to aid the passengers. At the end of June all Stensrud work came to a halt, having moved 6,164. However, 16 bodies remained in the old cemetery by families who "forever" did not want them moved.

The Oliver Co. is slowing down some of its work at the Sect. 16 and other mines. The Cambria will suspend operations and 100 men laid off. But entertainment continued with the coming of the Yankee Robinson Circus. All will have 5,000 reserved seats and see trained elephants, Egyptian dancers, and bands. It was June before the mailboxes arrived for the citizens to use for deposits of mail. 17 of them will be put around the town.

The Presbyterian Church no longer is having services, but the Mitchell M. E. Church will use the building for several months while renovations are taking place, including a pipe organ. The Union Station had a fire and just about burned down, but someone turned in a fire alarm. Notes the paper: "There should be a reward given for finding him." At Eagle Mills, many company buildings are for sale, including the sawmill itself, and the store. In Escanaba, 59 persons took tests to get a teacher's license. Only 11 got one. E. R. Gribble has come home here after doing diamond drill work up in Calgary, Canada. He was surprised that the town has 75,000 people. It is only nine years old. At the Bijou, the latest film is the Indianapolis 500-mile automobile race. It will also show a film based on the famous play, "Ten Nights in a Bar Room." The Adelphi Roller Rink now will be called "The Palace."

Up on Bluff Street, the children who went to the circus decided to do one themselves. They practiced stunts for several summer weeks and by August gave a 1 cent show, doing a high dive (so good they did it several times to applause), tightrope walking 14 feet high, and a singing minstrel group. The show was still going strong when the curfew went off and they had to take it all down and end the show. We hope they put all the garbage in a container. It is a new city requirement.

A new Michigan law is that no one can carry a concealed weapon without legal permission. The nice new Fire Station clock stopped when the winder went on vacation. Another circus came here and then went on to Michigamme. The Campbell Bros. Circus, unlike many of the shows which simply passed through Michigamme, stopped for a performance. There was a disaster. High winds blew down the tent and tore it to "smithereens." Ten elephants got so scared they easily pulled up their stakes and ran all over town, stopping by the shelter of Mrs. Sundstrom's Store. The keeper came and they followed him back to the train. Noted the paper: It was quite an entertaining day, and it didn't cost Michigamme people a cent.

Rosen Bros. and Klein closed their store at noon on Friday, and all the employees went to Presque Isle by automobile. Quite a treat! The CCI is meeting with the state regarding an improvement of Worker's Compensation with money going directly to the worker. In September, the County Fair is on, and this year it will feature an "Aeroplane." There is also this interesting news, that there will be the "first" Northern Michigan State Fair, and it will be in Escanaba. There is a new merged railroad now called the Munising, Marquette, and Southeast Railroad. The Miller Bros. have purchased a two-horse bailing machine for hay. Auto owners are asking for U.P. road maps. The county let bids for a road from Little Lake to Delta County.

The teacher gave Johnny a problem to do. How many tons of coal at $6.00 a ton would you get for $24.00? "About three tons," he replied. "That's not right," she said. "I know," said Johnny, "but they all do it." At the Negaunee Mine, there was another run of sand and water and work stopped for a few hours. President Taft came to Marquette. Large numbers from here went to see him. Woods workers this year will get from $26 to $35 a month plus room and board. We also note that Finnish women have complained that they cannot also go into the woods and help their husbands as they did in Finland. The logging companies have settled with them by letting them be cooks and workers at the logging camps.

Theaters are getting fancier fronts. The Bijou's is very ornate in the Sundberg Building, and now Mr. Rytkinen's new theater also has a recessed and ornate front. It is now October, and Mr. Ammonino has a pumpkin in his yard that weighs over 53 pounds. The Order of the Owls is open to new members. It has sick and accident benefits and free physician and burial expenses. Bounties were paid in October for four wolves killed in Tilden Township. There were mine paydays this week and the stores will be open every night. However, there were also mine layoffs in the paper of 200 men in three mines.

The Negaunee City band was formed by the merger of two bands, the Negaunee Light Infantry, and the Negaunee Concert Band. A few weeks later the "old" City Band also reorganized and chose Mr. Joe Violetta as the leader. The cemetery will build a new winter vault to hold 40 bodies. Skunks are numerous in town and are raiding hen houses. The Palace Roller Rink has hired a comedian and one evening he played the role of a man trying to skate for the first time. Going through to the Copper Country this week were seven full railroad carloads of grapes for wine-making. A report on the water of Teal Lake, 40 feet from the shore, shows it still to be "unsafe," with animal organic material in the water. Almost 40 Finns have left here in two groups, for Finland.

A company near Seney is selling land for those who wish to farm. None realize the harsh winters and many will leave. In early November, many people are washing windows. After all, they are already soaped. We hate to put in bad news, but this one may set a record. Negaunee's football team lost to Menominee, 81 to 0. The Mine Inspector's Report showed 32 deaths, and notes that the mines are working on rescue equipment such as gas masks. And next to Iron Street, at Pit #7 of the old Jackson Mine, the shaft house was removed.

At Christmastime the Salvation Army gave a dinner for the poor in Negaunee. Many called in names and contributed help. Wm. J. Maas died, who found the Maas Mine with his brother, George. "He was quiet and helped people in a charitable manner."

Here is a Street car on Iron Street coming toward Cyr and then on to Ishpeming. Photo from 1910. Photo from the Negaunee Historical Museum.

Three fine trucks arrive for use in Negaunee by the City. They paid for themselves over a period of a few years when compared with horse expenses. Photo from the Negaunee Historical Museum and C. Ruhanen.

The Marquette Milling Company opened up a factory at the old Peninsula Bottling Works on Gold Street, and, as many new businesses did, sold stocks. Some bought a hundred shares, but some far less. Anne Pynnonen bought two shares and paid $20.00 in July of 1919. Negaunee Historical Museum.

1912

An electrical problem caused by the shorting of wires of the Electric Railroad's direct current, and the city alternating current, in high winds, burned down the Negaunee State Bank and the Mqt. Telephone Co. upstairs. The very next week the telephone service was restored with the new switchboard being in the Sundberg Building. The safe of the bank survived the fire. The year started out cold as well. In the first two weeks there was only one morning above zero, at ten degrees.

Negaunee has adopted a new round seal with a drawing of the stump at the Jackson Mine, with the words: "Negaunee City 1873, Founded 1846 (1845?), Village in 1865." "Iron ore first discovered in the Upper Peninsula of Michigan (1844)," and "At Negaunee, in roots of this stump (tree?) in 1844 (1845?)." Charles Chalifoux is the last of the first settlers to die here. His home is one of the first half-dozen log homes which constituted the town. In the woods, lumbermen are now busy cutting hardwood which was passed up for pine in the past. Stewart Gaffney has died at 18 of tuberculosis. He planned to graduate this year. John Allen at the city light plant says his records show only one morning in January when the temperature was above zero.

We have the first mention of asphalt for street pavement on Iron Street. There is no mention of any wood blocks on the street. In early March, there is mention that Vermont Maple Syrup has arrived. This is amazing, noted the paper, as the trees in Vermont haven't been tapped yet. The S&I depot for passengers has been moved from the south end of Silver Street to the new railroad route and will be across the street from the Jackson School.

75. Presbyterian Church Ends

The Presbyterian Church, closed for the last four to five years, may be sold. It is an open secret that the School Board is interested in it. Active members feel that the re-opening is remote. Some feel the building could be a YMCA.

The plans for the new Negaunee State Bank will still be on the same triangle on the south side of Iron Street, but will have the entrance in the middle on Iron, not at the tip on Iron as before. The Negaunee basketball team went to Crystal Falls and won the game and then was treated to a dance with music by the high school orchestra. St. Paul's Church is having lenten service three nights a week, on Tuesday, Friday, and Sunday. The Case Street adult classes have 85 enrolled and English classes at three levels of reading. And finally, it looks like the eight-hour days are becoming common in the mines. Several have had it, but now the CCI and Oliver Companies will follow suit.

There are some highly unusual news items in the paper now and then. Does anyone know more about this one? "There is an aerial Ferry in England that carries 500 people and six vehicles. It has a height of 350 feet and is suspended by cables and moves using an electric motor." The summer professional baseball league is dropping Escanaba, and picking up Calumet. Gladstone wants to come in and see Marquette dropped. Marquette was not dropped. Mr. O'Donoghue has donated a large sterling silver cup to be given to the Negaunee baseball player with the highest batting average this summer. Lots of babies are being born. Eight are in the paper in two weeks of April. Elections in April were for a $35,000 bond issue for a new City building. Four counties in Michigan that had been dry, went wet.

1912

An electrical problem caused by the shorting of wires of the Electric Railroad's direct current, and the city alternating current, in high winds, burned down the Negaunee State Bank and the Mqt. Telephone Co. upstairs. The very next week the telephone service was restored with the new switchboard being in the Sundberg Building. The safe of the bank survived the fire. The year started out cold as well. In the first two weeks there was only one morning above zero, at ten degrees.

Negaunee has adopted a new round seal with a drawing of the stump at the Jackson Mine, with the words: "Negaunee City 1873, Founded 1846 (1845?), Village in 1865," " Iron ore first discovered in the Upper Peninsula of Michigan(1844)," and "At Negaunee, in roots of this stump (tree?) in 1844 (1845?)." Charles Chalifoux is the last of the first settlers to die here. His home is one of the first half-dozen log homes which constituted the town. In the woods, lumbermen are now busy cutting hardwood which was passed up for pine in the past. Stewart Gaffney has died at 18 of tuberculosis. He planned to graduate this year. John Allen at the City light plant says his records show only one morning in January when the temperature was above zero.

We have the first mention of asphalt for street pavement on Iron Street. There is no mention of any wood blocks on the street. In early March, there is mention that Vermont Maple Syrup has arrived. This is amazing, noted the paper, as the trees in Vermont haven't been tapped yet. The LS&I depot for passengers has been moved from the south end of Silver Street to the new railroad route and will be across the street from the Jackson School.

112   Also: Correction for p. 78 (2nd Column). "In them," should read "In the Mines"

This story comes from game warden, John Rough. Two men fishing at Morgan Pond before May 1 when the trout season opened, saw John Rough coming down the tracks, and left so fast that the game warden now had a nice pole and a can of worms. He waited there until midnight and by that time there were 27 men there waiting to begin fishing legally. Before leaving, he fed the worms to the fish. We can also tell you that at this time, there are still trout in Teal Lake. James Bennetts, 10, caught a 19-inch Brook Trout just short of three pounds. He caught it by the power house with his stick pole.

The city is beginning its road work and has purchased two new road wagons. We take it for granted they were what we now call a truck. The city officials visited several streets in Wisconsin and Illinois to see how well asphaltic macadam held up in heavy traffic, for consideration on Iron Street. Another old-timer died, Mr. Benjamin Williams, 93. He used a couple of teams to build the tram road to haul ore from the Jackson to Marquette. The county still has four Civil War veterans, E. C. Anthony, J. O'Brien, J. MacNeil, and Peter Trudell.

Mr. James Foley has sold his residence at Case and Teal Lake Ave, and is going to California. 40 people went to Lake Michigamme to the YMCA camp there. They made it in only five vehicles. Here are some auto names being purchased: Palmer-Singer, Lexington, DeTamble, and Kissel Kar. For those who have a dandelion problem, the newspaper notes that you can get rid of them by putting iron sulfate on your lawn. The Negauneesian is available. It is number six. Two theaters, the Bijou and the Ishpeming, have closed as it is summer, and mines are slow. Could be either reason. The first street address is in the Iron Herald, and that is that Charles Forell is opening an auto agency in the livery barn at 309 Iron Street. The first note of a carnival, other than merry-go-rounds,

came to town with rides called the Flying Bullet, and the Kansas Cyclone, and it will also have the largest Ferris wheel of any road show. In addition, Ringling Bros, the Hagenbeck-Wallace and Buffalo Bill's Wild West shows will all be here in the summer.

The L'Anse to Skanee road is being completed, and the road from Marquette to Big Bay is completed. Noted the Herald, "Marquette County has only to bring their road from Big Bay to the Baraga County Line." The paper also notes that it is a beautiful route and will "also open up the country to settlers." (This road, of course, was never completed.) The government is planning to coin a new three-cent coin. And in late June, the Negaunee baseball team is in the bottom of the league. There are several non-professional teams in town as well. Two are the Rexalls, and the Buffaloes. On Iron Street there are now 20 street lights, fastened on the Electric Railroad poles.

The Negaunee Mine also was forced to close for a few days because of a run of water and sand in the old workings. The Presbyterian Church will be purchased by the Board of Education. The Cleveland Park that is so popular now has a zoo. It holds two bears, three deer, three fox, 48 rabbits, a prairie dog, a woodchuck, and two crows. Drivers of automobiles are being told to drive on the right and blow their horns when crossing another street. A watchmaker on Iron Street, J. H. Sunne, has just returned from a purchase trip to Milwaukee and Chicago.

In July, Mr. Israel, who took a lot of photos of the town, published a book of his photos. It was recently republished by the Negaunee Historical Museum. In the autumn, Mr. Israel moved to Port Huron. Some visitors arrived here by auto from Chester, PA, and in August a group came from Terre Haute, Indiana, 800 miles with just one blow-out. After they left Marquette they were in Menominee in the afternoon.

At the cemetery, a four-inch pipe has been laid from Mud Lake (by today's US-41) for watering. It will also be tested for drinking. The county is now working on a road from Dexter to Humboldt. They are also working on a road from Gwinn to Palmer. In town, Lincoln Street is being extended east from Healy and 17 lots will be made south of Mr. Alexander Maitland's home on Main Street. George Maas is platting his land for 240 lots at $250. each. You will get a 10% discount in the first month of sales. Two girls were sent home to Ishpeming, age 15 and 16, for disorderly conduct.

And the wonderful news is that work in the mines begins to look better. The Hartford reopened after 18 months, and the Maas Mine will resume after a quiet two years. About 75 men will be hired now and increase to 300. The Volunteer Mine will reopen and hire 150 men. In town, stores have not yet begun using street numbers in their ads, but more telephone numbers are in use. The death of young Ruth Gribble, 15, after an illness of three years lists her address at 515 East Case Street. The large church bell in the Methodist Church has been given as a gift from Mrs. Sam Mitchell, it being formerly in the Presbyterian Church.

The State reports that Teal Lake is unsafe for drinking purposes and it is suggested that Silver Street spring water be used instead. The level of the lake is also receding and the city is looking out at the Cliff's Driveway. As many young people are getting post office boxes and keeping their parents in the dark, all mail to a family will now be delivered to the home. In September, the professional baseball league final results showed Negaunee in second place, after Ishpeming, short just one win. In

Michigamme, the people there are installing their first electric light plant. City employees, cleaning a fire alarm box, found two letters had been mailed in it.

Theodore Roosevelt came to Negaunee in early October of 1912. He was on the train on his way to Marquette, but since it stopped for a bit, he came out and gave a speech on the end car. It had a bit of humor as he didn't expect the train to stop so long, and he had to keep talking. Gladys Wiggins and another four-year-old girl were enticed to go with a man down an alley where he was seen by her mother who obtained the children, and was forced to bed by the ordeal. Police are working on the matter. Because of the Teal Lake problem, Mr. Tom Bashaw has signed a contract to provide the schools with spring water, 80 gallons a day.

A 21 x 24 foot chapel is being built at the cemetery. The library has received a series of books for home study in water, plumbing, sewer, hot air heating, masonry, pipe work, bookkeeping, and more. The Mine Inspector's Report shows only 15 deaths, and only three from Negaunee. Safety equipment enabled the lowest deaths in County mine records. The School Board found that the Presbyterian Church will have to be torn down as it falls short of the required fireproofing. The church can dispose of the pipe organ. (No record of it's demise has been found.) There will be no bowling alley this year as there seems to be little interest.

In December, another cave-in on the lower levels of the Negaunee Mine stopped workage for a few days. There is little snow and wagons are still in use. Lots of toys appearing and there is a Christmas ad already on December 6th. The City Band will play for a dance on Christmas Eve.

1913

Story hours for Children will begin at the Library on Saturday, January 18, at 2:30. They met with great success. The Michigan Legislature has a bill pending requiring paydays twice a month. A six-frame cartoon is found in the newspaper. Rasmussen's Store has shoes for $2.35, overcoats for $14.85, boys' winter caps 29 cents, heavy wool sweaters $1.85, and ladies' petticoats for $1.69. A mask party at the Palace had one requirement. You must be able to roller skate. The Swedish Lutheran Church gave a surprise party for the successful work of their pastor and wife. Rev. C. E. Lindquist received a nice purse with $30.00 in it and his wife received a fine silver plate.

Electricity may be used to light the road and homes between here and Marquette. The city of Escanaba bought some nice looking voting machines, only to find out they are worthless. Wm. Tresidder and Wm. Johnson have won several prizes at the Copper Country Poultry Show. Mr. John Nadeau died and left a record of 18 children from three marriages. The city now can use Teal Lake water again for drinking, including the schools. A new intake pipe into the lake has been installed. And this note may be the last thing we hear about Jack Carkeek as he has been sentenced to six months in prison for fixing wrestling matches in the USA and in Europe.

76. Main Street Abandoned

The CCI has met with Main Street property owners to discuss the abandoning of part of the street, but the city owns the street and has the right to reopen it after the mining has ended. The new road will leave Main at the (old) Baldwin Kiln Road and go NE for 500 feet along side Corbit's 2nd addition and then a 1000 foot road will be built back to Main by the old Maynard Gauthier farm.

Another new news item is the recording of the "First Hockey Game to be ever witnessed in Negaunee." It was at the outdoor city rink and Mqt. got five goals and Negaunee only two. Our players were Wm. Pelto, Bert Rosevear, Charles Tall, Aloysius LaMere, Roy Cullis, John Cullis, and Wm Davey. They played another game the next week of February. The new Iron Street lights give off a thousand watts each and it looks like a "Great White Way." The Ski Tournament took place with two 143 foot jumps and Mr. Hendrickson doing his somersault off the bump. February was very cold. Only two days got above zero.

Sports could get out of hand in 1913. Houghton came to Negaunee for a basketball game. The referee allowed about everything. Houghton won, 31 to 15, mostly by tripping. The game was rough and there was one fight on the floor. Many railroad workers are still getting injured when the cars bang together suddenly and they fall off ladders and roofs to the tracks. There is a new automobile dealership open and Kirkwood's are selling the Michigan "40". It has four forward speeds, electric lights, 22 coats of paint, and oversize tires, for $1,585. In the April elections, women were still trying to get the right to vote. Less luck this time than before. There was also another blasting cap incident when Edwin Miller, eight, found one at the Cambria Mine. He poked at it and it went off. His hands and face were badly lacerated, but his eyes were saved.

Fred Braastad's Negaunee store is being remodeled and he has bought out the stock of the MacDonald's and Laughlin's stores. The US is placing new paper currency in circulation, and it is smaller than the current bills.

77. A New City Hall

In May, plans are drawn for a new city hall. It will be where the present one currently is, and although it was hoped to use some walls of the current building, the foundation was found to be poor, and the entire building was removed. The library will go to the high school, and other offices moved as well. One thing will come to the New City Hall. It will be the small clock in the fire hall tower. It was discovered that the works were made to handle larger hands, and the new face will have a diameter of seven feet. It will be all stone and brick with the entrance on Silver Street. There will be a room for the children's hour in the basement. The jail will also be there. The newspaper notes that "It is hoped to serve for the next half century to come." (In 2013 it will have served for 100 years.)

Edward Mayle carried off a 4-year-old girl, Celest Hill to an alley and was taking liberties when she screamed and her friends went and got her mother and he was beaten up by Mr. Hill and then arrested by police. Julius Mattson stayed at the Montreal House and paid no bill. He now will room with Undersheriff Sam Bennett. Negaunee and Ishpeming each have 11 graduates from the Normal in Marquette. Michigamme High School has seven graduates, and nine graduated at Republic.. Five Socialist speakers are speaking in the streets and were arrested one by one. The last was a lady. More and more people are traveling and the news notices now fill over two full columns.

The Street RR evidently never was moved as it is now still ending at Cyr and Iron St. until the Iron St. paving is completed. Many Italians are in town for a banquet and a day at Cleveland Park. James Pickands and Co. is selling coal and wood. The Street RR has a policeman at the Cleveland Grove to keep out unwanted characters. A new law will allow widowed mothers to receive aid for their children. Holzhay is now released from prison at age 47. Mrs. Arland is selling her millinery business. Merchants will close on Wednesday afternoons in July and August. John Brown, in the Cambria Mine for 28 hours until found, in 1909, died. He never fully recovered. The Singing Indians from Oklahoma were at the Star Theater with a good audience. Richard Nesbit has put in a mechanical orchestra with a piano, mandolin, and other instruments.

The Main Street bypass is completed and rejoins Main Street near the old cemetery entrance. The city is filling in the marsh between Teal Lake Bluff and east of Teal Lake Avenue. The area will be used as a playground and have play equipment. The Fair in Marquette will feature motorcycle racing. Businessmen are upset that the post office may move to Jackson Street from Iron St. Mr. Baptiste Barassa has some pumpkins that are over 5 feet around. The Negaunee football team in September lost to Ishpeming, 64 to 0.

78. The First Homes Moved

As the new Maas Mine Road to Mqt. is open, the CCI will move 15 homes on Main Street up to the old cemeteries. The Star Theater is putting in a balcony to hold 150 more people. Mr. Honkavaara and Nesbitt went to Indiana by auto for 1300 miles without a breakdown or a delay. They came home from Milwaukee in one day.

In October, Teal Lake is again tested unsafe, and the city must provide water to schools. The city will begin to filter the water from the lake. The Bijou Theater is now called the Royal. In his saloon, Julius Johnson caught 11 rats one night and 9 the next. The mine report for 1913 shows 48 mines are operating and 5,700 men are at work on the Marquette Range. 16 men died, 10 of which were Finns. 64 loss-of-time accidents. South of Republic, at Floodwood, a man killed 17 wolves for the bounties. Several people in town witnessed a cat catch a rat by the neck, but was unable to kill it, so it dragged it over to a pail of water and held its head under until it died.

This interesting note regarding Negaunee's water: The state says that a chlorination plant will have to be installed. The state then added that, along with Ishpeming, they "might join forces and combine in a joint system." The November 8th paper notes that the weather is so nice that the farmers are busy plowing their fields. The city of Negaunee has printed the new charter in the paper for people to read and eventually vote on. John Reichel has a large amount of electrical fixtures for sale at 323 Iron St. Your 1913 Red Cross Seals should be in the mail.

The year ends with the Negaunee and Maas mines hiring a total of 100 men. A trench from Horseshoe Lake down to a sump has been completed to get water to the cemetery. Mrs. Arland sold her shop and is moving to Los Angeles. The city will now be buying its electricity from the CCI as they have it to sell, and your lights should not flicker any more.

Here are Hugo Honkavaara and the future Judge Davidson on their homemade car using a bike engine and wheel behind. Courtesy of Avery Color Studio's: U.P Auto History by Perly.

Courtesy of Avery Color Studios

Hugo and Judge Davidson headed for Soap Box Derby
Top speed 35 mph

1914

Sam Collins knows it is again a very warm winter. There is only snow in the shadiest spots. Automobiles are in use and the dust flies just like in the summer. He found that the winter of 1877-78 was also quite warm. Ore shipments were down at the two Marquette docks, and the number of ships this year was 536 versus 596 a year ago. The copper strike in the Keweenaw continues and 175 strike breakers passed through this week, picking up 525 ham and egg sandwiches and 25 gallons of coffee from Andrew Erickson's Restaurant. A week later, seven more carloads of strike breakers came through.

The city is asking citizens to vote "Yes" on a new charter as it is needed for election of officers and the running of the city. Two men died at the Negaunee Mine from a fire and they were overcome by gas. Capt. John Barrett had taken off his mask for some reason, and John Beebe did not have one. The people in the Copper Country are still waiting for a way to get to Marquette County, and businessmen in Crystal Falls can help with the connection by raising $6,000 to get a road from there towards Republic. People have quit roller skating and all of the Palace equipment has been shipped to Sturgis, Michigan. Bowling, however, is doing well.

The city charter did not pass: 351 against, 321 for. Three Negaunee stores also closed during the winter. In Humboldt, a government thermometer registered -45 on February 12. A quick run of quicksand in the Maas Mine killed John Juhula, 36, leaving a wife and six children. Tom Curtis discovered the problem and more would have died if he had not entered the working area and warned the men. In schools, the Palmer method of penmanship was introduced and is having great success. All the children were asked to write a letter to the Superintendent. He said that many letters looked better than writing he gets on teacher applications. The state has stated that charging for auto licenses by horsepower is illegal, and all those who have paid over $3.00 will get some money back. Mr. George McCann, who opened a fine restaurant a year ago, has closed it saying, "Negaunee is a poor restaurant town."

The seven-foot tower clock which was to set all the other city hall clocks, will not do so, and save the city $5,000. The state weight and measure man is in the area and is finding numerous false weights and measures. Several were destroyed and others will be rechecked. A large storm came to town and carried in a seagull. He was given some food at the Breitung and left the next morning. A new funeral director is here, Mr. Wolley, and will be located in the new Swanson Furniture Store. Levine's have purchased the large Neely Block and are installing a large steel beam to eliminate the poles on the first floor.

In May, for the new City Hall, the city will purchase three lots towards Teal Lake Avenue, removing a home and a warehouse. Some offices will be temporarily in the upstairs of the fire hall. Fishing season is also underway, and the limit for trout is now only 35 fish a day, or 50 in possession, and all at least seven inches. A report on the strikers families in the Keweenaw told of how 13 homes were visited and there was not a total of $10.00 of food, with one family having eaten only potatoes for the last three weeks. Dr. Andrus has bought a new Franklin auto. It is air cooled and can run all winter.

Iron sales do not look good this summer. Republic has let over 50 go, and cut down the work week. The CCI Marquette Furnace will go out of blast. Mr. L. Rinne, however, who sold his jewelry store here ten years ago and moved to Illinois, has returned to live here again and bought the Wentila Jewelry Store. The remaining members of the Presbyterian Church have elected officers, in order to maintain their $10,000 and are able to build another church in the next 20 years.

The city is now to tearing down the present city hall. The building was built on pilings and a new basement has to be put in. Quite a few people are buying their water through Eugene Robert's pure spring water home delivery company. An ad for the Miami Publishing Co. advertises a 320 page "Sexual Knowledge" book for $1.00. It "will be sent in a plain wrapper." It is July 4th, and "Homecoming" time again in town. There is electric night illumination, a Children's Parade of 1000 children, dressed in costumes, and marching like soldiers. The parade was almost a mile long. All the school teachers were given a free trip to Marquette and a meal out as a thank-you. All visitors from beyond the county signed in and there were 250 names.

The Chautauqua group will be at Cleveland Park with educational talks, and recreation not usually available here to the public. In August, the city library, supported by penal fines of the justice and circuit courts of the county, received a nice check of $833.84. The same week, there was also a frost, but a light one. It is a bad summer for the Negaunee Professional Baseball team. They are in the basement of four teams. Escanaba is first. At the Royal Theater, the series of "The Perils of Pauline" are being shown each week. 3,000 carnations are being grown by Frank Ashleman at his new greenhouse on Merry Street.

In the sporting world of Mother Nature, there have been several arrests. One man was catching deer with a trap, another fishing by using dynamite, and a third hunting with lights. The trap had one-inch teeth, and two men standing on it could not open it. In Ishpeming, Mr. Trebilcock is tearing down the 24 buildings of the Ropes Gold Co. as it comes to an end. In Negaunee, the summer work at the two rock crushers came to an end at the new quarry. It is at the west end of the Jackson Mine and a few rods north of the county highway to Ishpeming. The asphalt paving size rock was taken to Teal Lake Avenue in DSS&A, 50-ton, cars.

Since there is no call for iron ore, 500 men have been dispensed with on the Marquette and Gwinn ranges. The newspaper notes that the European War is primarily responsible for the present conditions. In October, the Negaunee Band gave a two-hour concert on Iron St. to a large crowd. The clock was removed from the fire hall, and the fire bell will be installed there. Cows are still a problem. We have skipped some complaints in this book, but a herd of them are wandering around a section of Main Street. Some residents are thinking of hunting "Big Game."

The October Mine Report for the year shows 22 deaths, 345 one to ten day injuries, and 200 that missed over ten days of work. 33 mines are operating, 82 are abandoned and the total men employed during the year, 4,969. There are several cases of diphtheria and one death so far of 14 year-old Julia Terris. People are donating toys for children in Europe who will not have Christmas because of the war. They will go there on the "Christmas Ship." The Western Federation who have led the Copper Country Strike, have had four of their leaders arrested for thee men killed last December and will go to trial in Marquette. Ben Neely has sold his hardware store at 440-442 Iron Street and will winter in a milder climate. John Thompson owns a tract of land in the Huron Mountains and

it contains vanadium. It is in T52-R29 and in Section 36. Housing for men is going up, and a shaft going down. There is manganese in the area as well.

The newspaper made a bad error in November, saying that Amelia Bogetto, seven, had died from diphtheria. It was not true. She was not even sick, and furthermore, she showed up at the Iron Herald office to prove it. There is a "foot and mouth" disease sweeping the country in November and killing herds of deer. It has not swept through here yet. Lots of hunters came across the straits again this year, 3,562 of them. In December, it was another nice month, with ice on Teal Lake, but no snow. Perfectly ideal for skaters. Phone 180J if you can help the Salvation Army feed the needy with Christmas dinners. The schools have placed 25 trees in classrooms, the eighth grade will present a play on George Washington at the High School assembly room, and the Board of Education will give each child some candy and an orange.

Some people made money raising and shipping Christmas trees to Chicago. 200,000 have gone there from the U.P. The Supt. of the Michigamme Schools was arrested, then released, for disciplining a child. The School Board told the judge that was the exact reason why he was hired. Not much about Christmas in the paper. There were only four ads mentioning the holiday in some way, but a nice Christmas poem on the front page of the December 25 paper.

The size of paper money changed in this time period.

1915

Four pages of tax sales are listed in the Iron Herald in January. Evidently the hard times are causing many to fall behind on their taxes. Mrs. Kruse, who recently lost her husband who was a member of the Modern Woodmen, received his insurance check for $2,000. The Lowenstein Building is going up on Iron Street and will extend all the way to Jackson Street in the rear. People can read an article on "How to Save Electricity." One interesting suggestion is to wipe fingerprints and dust off the bulb. It won't get so hot and will last longer. The girls at Negaunee High have formed a basketball team. The girls lost their first basketball game, 40 to nothing. The boys lost even worse at 71 to 17. They both played at L'Anse. The Negaunee boys also have a hockey team and they lost to Ishpeming, but only 1-0.

At the end of January, a huge fire on Iron Street burned down five buildings. Some were brick, but the brick did not extend up into the roof joices, so the fire went from building to building until reaching the Sher Store at 518 West Iron that had a full-height brick partition.

79. Destitute Families in Winter

The city is seeing if they cannot put some destitute families to work. There is hardly any snow to shovel so not many have been hired so far. The paper asks its readers to inquire in a block or so of their homes to find some people that can probably use their help. Ward Committees have been formed by the women to do relief work. They have concluded that families without a husband or father need charity, with a husband or father, they need work. Children need clothing, and provisions. An ad was put in the paper as to where to go for aid. They are only taking applications now and are weeding out undeserving applicants. Many

of the needy are not coming forward and so help will be given indirectly. Some have wood, but bare cupboards, others have flour but little wood. Some need only shoes. The city received word that the Marquette Pioneer Furnace will hire men for a few months to cut wood. By the middle of February, 31 families had been given aid.

The Breitung House, as usual is keeping up-to-date and remodeling. Mr. Seass, long-time manager, has put steam heat and hot and cold water in the 22 rooms on the second floor. The Negaunee girls played L'Anse a second time after losing a few other games, and lost by only two points, 29 to 27. In bowling, Negaunee men beat the Ishpeming men, 2264 to 2246. In a third match with L'Anse, Negaunee won soundly at 26 to 5. Catherine MacDonald was the star. More men went to work in March when the Cambria and Hartford near Teal Lake went from three days a week back to five and more men may join the 25 working. The city found some work for men by replacing the wooden water main from Teal Lake to the city with new iron pipe. A water bond passed, 364 to 152. The C&NW is going to replace all of its iron rails, which often break, with new steel ones, costing three million dollars.

The city librarian is ready to move from the high school to her new rooms in the new City Hall. The newspaper notes that Ishpeming and Marquette provide their librarians help, but in Negaunee the librarian is expected to do the work alone and notes that there is even more disparity in terms of librarian salaries. 9,000 books need to be moved. Geologists are not too good at determining altitudes. They conclude at this time that Frost Junction in Houghton County is the highest point in Michigan at 1,407 feet. The ladies' spring skirts are only about six inches above the ground. There will be a problem of their sweeping the sidewalks of spit from miners chewing snuff. In April, St. Paul's Church was robbed, along with some other places. Authorities later found the men and lots of loot, but no altar items from the church. They had thrown them in a river after they discovered they weren't solid gold. Smallpox is back with some homes quarantined. A new trout fishing rule came into being. Streams publicized as having been planted with trout cannot be fished for four years. All the automobiles are coming out of their garages and the editor writes: "We certainly have a bunch of autos now." Michigan has 900,000 licensed autos and 6,151 licensed motorcycles.

On Memorial Day, the Civil War Veterans and the Women's Relief Corps will visit graves. Autos will bring people to the cemetery and the Negaunee Band will play. Mrs. Lenore Klein sent some presents on the Christmas Ship to Europe and has received a nice thank you note from Greece on behalf of the baby that received the blanket. In June, the #10 Negauneesian went on sale for 75 cents. It has more photos. The Negaunee Alumni Assoc. held its annual meeting to prepare the program for this year's 20 graduates. There were 59 of them as freshmen. There was another death at "Dead Man's Curve" just west of Morgan Heights. Cement continues to be in even wider use. The Lucas Bros. Livery Stable has put in a nice cement floor. Al Sundeen on East Case St. has built a fine cement floor in his dairy barn for ten cows. An ad in the paper caught a lot of attention, reading, "The War is Hopeless." It was about dandelions.

There was quite a commotion in town when the Weight and Measures Man came and found many problem scales among the 388 he checked. 19 were closed down immediately, 60 shut down until repaired and 56 were re-regulated and continued in service. On June 22 there was a very heavy frost---and on the longest day of the year. The paper notes that the sun came up at 4:13, and went down at 7:24. (Yes, you can read that again)

80. Good Times Again

The good news was that the Oliver Co. will hire several hundred men at the Prince of Wales and the Section 16 Mine. The Maas Mine which has been idle since October, in July hired 200 men. Soon ore shipments were 500,000 tons above last year's figures for the same period. The CCI, which had reduced wages, increased them on August 1. The Republic Iron and Steel Co. raised their wages 10% as well, and so did the Mary Charlotte and the Breitung Hematite. Those two mines also went on three shifts a day with 400 men.

The Howard Clock Co. has returned the workings of the tower clock and lengthened the pinions to adapt the new seven foot hands on the four sides of the clock that will be installed at the City Hall. New furniture and equipment is also arriving. In July, there were two heavy frosts on July 24 and 25. A bounty of five cents is now being paid by the state for rats. And travel by car is going well. Many coming here to visit from out-of-state, and Dr. Andrus and his family, with a chauffeur made a trip out east to the Allegheny Mountains for three weeks. Not one puncture or mechanical problem for 3000 miles. It was a radiator-less Franklin and they said they even had some paved highways.

On August 18, like June and July, there was another heavy white frost and another the next week. Over in Munising they are recalling the mayor. At the Negaunee Montreal House, one can go in and see a six-foot live pine snake, caught crossing the highway between Ishpeming and Negaunee. Three men escaped from the Marquette Prison and jumped off a train in Negaunee and did some robbing at night with men in pursuit. It was quite an evening, as they had split up, but all were caught and the police got to split up the $50.00 reward.

81. Final Bodies Moved

The final few remaining 13 bodies left in the old cemetery by six families wishing to remain there will be moved anyway and the families were notified of the same. Families may move them themselves if they so wish. The area will be caving ground because of the mining underground. September came, and there was another frost. The public was notified that courts have given bicycles the same rights as automobiles. 41 Negaunee graduates are now attending 11 different colleges. The Marquette Prison is built for 312 prisoners. It now has 383. The Star Theater is showing films of Mary Pickford and John Barrymore. Florence, Wisconsin, has a street named "Negaunee."

Mr. Wells, the weight's man has tagged the city scales until repairs are made. A few weeks later they were found unrepairable, so they sat for quite a while. The scale people have one for the city and are just waiting for the word to install it. The war is also getting closer to home. Mr. Ed. N. Breitung has a French steamer that was torpedoed and sunk by a German submarine on the Atlantic Ocean. It was a bad deer season and it was estimated that less than half of the hunters got the one deer they were allowed.

There were three pages of Christmas ads. Several Christmas ads used "X-mas." Fred Braastad's advertised toys and prices: Electric aeroplanes for $7.50, repeating air rifles for $2.75, and a girl's electric range for $8.00.

This is a photo of Rev. James Stanaway who started Sunday Schools all over the Upper Peninsula and Northern Wisconsin and lived in Negaunee with his family.

1916

Night School began in January, and 72 people showed up. Most were men of middle age. CCI wages and several other companies raised wages again for February 1. The Ishpeming Ski Club is quickly building a temporary ski scaffold at Cleveland Park for a tournament on Washington's birthday. The Negaunee chief of police is removing slot machines which seem to be numerous in pool halls, saloons, and even drug stores and a candy store. Two more fires on Iron Street, but the fire hydrants did their job and only the Kuhlman and the Brown Buildings burned down. The Negaunee merchants "Dog Derby" was a "Howling" success this year with over 50 dogs participating with their masters. New prizes went to the smallest boy and the largest dog. Sam Richardson won the racing prize.

The library is having its children's hour in their basement room at the library, and in March, 512 children attended. The final gathering in April was upstairs and was a "Victrola Concert" of recorded music. When that entertainment ended, a new entertainment began. This was the summer of visiting circuses and carnivals, four of them. Cole Bros. came in May, The Wild West and United Railroad Shows in June. They also went to Republic and Michigamme. In July, the popular Barkoot Shows came with carnival rides and good weather and an excellent band that many came out to hear. In August, the Parker's Greatest Shows came on two trainloads of double cars and a big midway and rides galore.

Many other exciting things took place in the summer. A modern cement mixer put in the Lowenstein foundation. The CCI gave another raise on May 1. The number of school age children in Negaunee increased by 144 to 3,075. And many came to the M. E. Church to hear the Alabama Colored Jubilee Singers in concert for only twenty-five cents. The CCI has purchased four ore carriers and one is now named "Negaunee." It is a 6,300 "tonner."

Here is an interesting question: "Why do horses eat grass walking backwards, and cows, going forward?" No answer given. Three plans are given by the county for removing the "dead man's curve" near Morgan. The Marquette County Fair is awarding a pony to the child who sells the most tickets. In September, after a year and a half, the weigh master is still waiting for Negaunee to install a new scale.

The steel business, states the paper, remains strong in September. There was a cave-in at the Mary Charlotte and they lost their ore pile. It was 150 feet long and 75 feet wide and also affected the Rolling Mill. Work continues but that area has been abandoned underground. In the autumn there are lectures for the public, a chicken pie supper, weddings, Halloween parties, basketball, and election candidates are busy campaigning. There are a lot of ads, also, for voting "yes" or "no" on Prohibition. With prosperity again, the newspaper now is usually eight pages or more every week.

December saw the CCI give its fourth raise of the year, another ten percent one. The salesmen of the Overland cars were invited to Toledo to see them being made. Jeeps are still made there. The city bandstand was moved to the empty lot on the high side of the city hall. The former place will now be used for the Community Christmas Tree. The newspaper had a large, grand, "Greetings of the Season" on the front page, and the now famous Christmas poem. People had money to travel and the railroads had heavy Christmas traffic.

People enjoy southern minstrels so much when they come to town, that the local citizens

dressed up and had their own show. It was a success with several hundred people crowding MacDonald's Opera House.

This is another cave-in at the Negaunee Mine which almost took the No. 1 Shaft, and did take at least one other building down to the underground. It later stabilized and formed a flat bottom.

Author Robert D. Dobson, and his wife Ethel
Alaskan Pipeline, 2009

Here is a photo of an electric mine motor that replaced trammers, or mules, in 1916, taking ore to the shaft. Top and bottom photos from the Negaunee Historical Museum.

1917

Twenty-one young single Negaunee teachers returned to their teaching jobs in other towns after spending the holiday's here at home. The Michigan Teachers Fund was declared legal and schools will start to remove funds from teachers' salaries for their pensions. The 1915 United Charities of Negaunee has disbanded. The remaining $413.13 was turned over to the Women's Club for their social work. Capt. Charles Doney died because a piece of rock would not come down a chute. Some dynamite was placed in it and lit. The piece let go on its own, however, and came down the chute and exploded at the bottom where he was. He was 35 and left a wife and five children.

82. War Seems a Reality.

It seems inevitable that the United States will enter the World War. Even children are encouraged to be part of a Savings Bank System where they can add money to their own account once a week and teachers will do the bookkeeping. 678 Negaunee children are participating. In February, the paper notes: "Everyone is hoping that war does not ensue." It was just a few weeks later, however, when the draft was begun and enrolled all men between the ages of 21 and 30. The Red Cross was organized in Negaunee, and citizens were cautioned to begin using food supplies sparingly.

Life went on as usual in many ways. Snow had come late this winter and frost is being found down in the ground to six and eight feet. 31 have already had pipes unfrozen with the town's new electric ice-thawing machine. A Boston city girls' team came to town and played the high school boys' basketball team. It was a difficult game for our boys, but they won, 36 to 28. In late April, the first automobile was able to drive the melted highway from Marquette to Ishpeming. The medical specialists, United Doctors, were in Ishpeming for a few days to see those with chronic diseases. They offered their services free of charge, but ladies had to be accompanied by their husbands and children with their parents. At the Ishpeming Theater the film, "Birth of a Nation," was shown. There were 18,000 people in it, 3,000 horses, and took eight months and a half million dollars to produce. On May 11th there was still ice on Teal Lake. It finally left on the 13th.

By May the war effort is in full swing. Many people are putting in gardens, and in June, the first big send-off occurred when seven Negaunee men went to Milwaukee to join the Navy. A total of 116 local men have enrolled. 88 have responded in Ishpeming. Even at Turin, the town that has disappeared, 17 men have registered.

The great numbers of automobiles have made it dangerous for people getting off of street cars in the middle of the streets. The high school has a number of Negauneesians for sale for seventy-five cents. Three Italian Societies have joined together at the Jackson Park picnic, for sports and with a band, after a morning parade. John Honkavaara luckily survived with only a broken thigh when he collided with a car on his motorcycle. The cycle was demolished and he was thrown right over the car. Another cyclist did die when he hit a wagon without lights during a dark night. The city would like to abandon the part of Silver Street in front of the city hall and develop a little park there. (July 13, 1917) The street extension was just built in front of the City Hall upon its completion in 1915. It evidently never happened until 2008.

The Red Cross is busy getting highly organized to support the war. There are now several

124

working committees, including First Aid, Hospital Garments, Surgical Dressings, Supplies for Fighting Men, Hospital Supplies, and Purchasing, and Shipping. People have donated $12,000. so far and a dance is being sponsored to raise more funds. Five more men left in July for the Naval Training Station in Illinois, and they were escorted to the train station by the Boy Scouts and the City Band. Another new war group that has formed is the "Four-Minute Men." They will go wherever possible and educate the citizens on the needs of the nation in this war.

The first city truck has been received by the Board of Public Works. It is a five-ton one, and will replace a team of horses that were costing the city a thousand dollars a month. The truck should pay for itself in one year. Those with horses yet are asked to construct manure boxes that will not leak into alleys and streets. Some mines are having trouble selling iron ore that is not of Bessemer Furnace quality. This has caused 110 men to be laid off at the Rolling Mill Mine. Some lads in August put little bicycle motors on their four-wheel carts they had made and drove around like cars with a fifth pusher-wheel at the back.

The governor is sending the National Guard men for war training. He tells the county here that the mines will remain safe without them for now. The Red Cross has three teams of 12 workers each volunteering to make surgical dressings. There are two long columns in the newspaper in July of names of men registered here for the draft. Things are so bad in Germany that people are reverting to wearing wooden shoes. In September there is a movement to supply the troops with cigarettes.

Hawks and owls are worth a fifty-cent bounty from the state. The Four-Minute Men will give their first talk at the Star Theater. The

Marquette County Historical Society was formed, and Mr. E. C. Anthony is on the Board. A Negaunee branch Historical Committee was soon formed as well. On the week of September 7 there was some snow and a frost, but there was little garden damage. Road progress continued when the first 2.5 miles of tarred pavement was made on the Negaunee-Marquette road by Morgan Heights. The Odd-Fellows Hall at Iron and Tobin will be remodeled for a possible movie theater downstairs. As was usual, the Jewish store owners, Levine, Lowenstein, Klein, and Sher, will close at sundown on Tuesday to the same on Wednesday for "Atonement Day." Rev. James Stanaway's wife passed away, leaving five sons, of which William and Norman reside in Negaunee. Mine employment continues to hold up with 500 more men at work this year than last on the Marquette Range, and new mines are at Palmer, Humboldt, and North Lake near Ishpeming.

The nation has been raising money now through "Liberty Loans." The town has been asked to raise $305,000. in buying bonds on a second push. Negaunee did even better on this 2nd loan, raising $28,000. more than the expected quota. A notice appeared in the paper for volunteers to join the army and enlist as clerks, typists, teamsters, etc. The ability to enlist will end at the end of the current year. The State Police here in Negaunee, called then the State Constabulary, have been sent to Ingalls to protect the largest power dam in the area. A third group of draft names of those leaving for the war appeared in the November 16, 1917, Iron Herald. The Red Cross soon had 2,421 members here volunteering to assist the war effort.

There are many cars now, and so many accidents that we are not able to stop and report many. One man was run over on the road to Marquette, and Mrs. Frank Roberts had her carriage struck by the city auto-truck and the horse had to be killed. Francis Kutchie, seven, almost was run over by a car on Peck Street, and may only lose an ear. These were in a one month period of October to November.

Many Finns went to Duluth to vote against the IWW whose strikes are not appreciated, and to oppose the invasion of Finland by Germany. John Schwartz celebrated his 80th birthday and told the news editor of how the mules hauled ore from the Jackson Mine to the Forge and then to Marquette on a tram road and that there were only a few homes "clustered about the Jackson Mine," and a few on what became the area of the Pioneer Furnace. They first lived in a log house on Jackson Hill. There were lots of large Christmas advertisements two weeks before Christmas, and a notice from the Owls that they were having a Christmas party for children "where the prospects for a Merry Christmas were not overly bright."

Here are copies of some checks found at the Negaunee History Museum of Negaunee Mines long past.

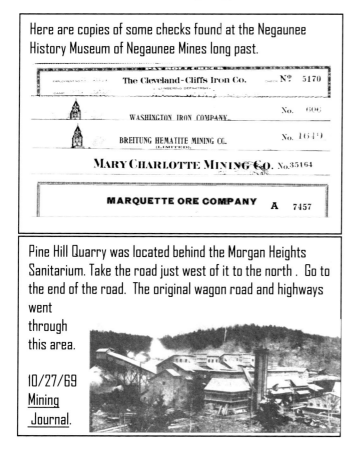

The Cleveland-Cliffs Iron Co.   Nº 5170
— LUMBERING DEPARTMENT —

WASHINGTON IRON COMPANY   No. 600

BREITUNG HEMATITE MINING CO. (LIMITED).   No. 1649

MARY CHARLOTTE MINING CO. No. 35164

MARQUETTE ORE COMPANY   A   7457

Pine Hill Quarry was located behind the Morgan Heights Sanitarium. Take the road just west of it to the north. Go to the end of the road. The original wagon road and highways went through this area.

10/27/69 Mining Journal.

1918

Grocery stores must list all of their inventories for the government as the amount of food stocks in the country is sought. In January, Thursday and Sunday nights were decreed lightless. The Friday Iron Herald notes "Last evening was well observed and electricity will be saved as well as fuel." (Fuel used to keep city power plants running.) The government is also now in control of all railroads and two trains of the CM&St.P. have already been cancelled. Churches are now allowed to be open only nine hours a week. Theaters must close on Tuesday and only be open for six hours on other days. Some stores can only be open until noon on Mondays, and saloons must be closed all of that day. People must begin raising poultry. Negaunee girls, 8 to 12, are helping the Red Cross in a big way by cutting materials for pillows out of scraps.

Myrtle Mitchell, daughter of Mrs. Sam Mitchell has wed the Bishop of the Protestant Episcopal Diocese in Oregon. Mr. Mather of the CCI will help students fill their "Stamp Cards." He will contribute stamps 1, 9, and 16 if students buy the rest to help the war effort. He will personally pay 75 cents of each $4.00 card. Negaunee people who are not yet vaccinated for small-pox should do so immediately. There are currently seven cases in Marquette. And then in February, the Health Department announced for the first time that Teal Lake water is safe to drink, right from the lake, and tastes better when cold.

A bad fire occurred at the Opera House from a stove in a hallway and did much damage to Fred Braastad's store and groceries downstairs. The Opera House has lately been used by the Negaunee Workman's Club. Seven poker players were discovered and arrested at the Burke Building and each fined $50.00. The worst part was that the names were printed in the paper. Otis Rule will open a motion picture house in the Odd Fellow's Block, and be called the "Liberty Theater." There will be vaudeville on occasion.

There was a cave-in near the Maas Mine between the new and old county roads. An area of about a hundred square yards dropped into the first level of the mine. Luckily, men were near the shaft, waiting to come up to surface for their lunch. A bulkhead will be built in the drift to keep quicksand from entering the rest of the mine. Nancy Mall, who with her husband, worked and lived in the winters at the Teal Lake "White House," died in March. She grew up at the Jackson Forge.

The war concentration is now on food and a large "Food Show" was given at the high school for preserving food and the best foods to be eating. The "Kenosha Plan" has been suggested here for raising money in Negaunee. It is a single effort to replace many small efforts, and then it is decided where the money should go. The names were printed of 26 more local men to be in the next draft in April. In May, the list was 35 of the total of 120 men from the area. The government will begin to call on people to see if they are hoarding flour. Families are only allowed one 24 lb. sack of flour. Citizens are to report anyone disobeying the current food regulations. A notice for more volunteers was made for making Red Cross dressings. The 3rd Liberty Loan drive is on and Mqt. County is expected to invest $1,542,000. to buy their share.

83. Liquor Sales End Here

The paper announced that in the State of Michigan, liquor sales will end on Tuesday, April 30th. Negaunee has 18 "thirsty-parlors" that will close their doors or sell other items. It was soon found that some temperance beer being sold was actually regular beer, or close to it.

There was a surface cave-in at the Breitung Mine and a new railroad spur into the property will replace the present one. Trout fishing did not open until May 15 in the U. P. It is recommended that all plant a garden and think about food supplies for winter. And finally, in 1918, a road has been built between L'Anse and Baraga and one can now travel for the first time, the route between the Copper Country and here, without using a boat. The Negaunee Italians in June formed The Lega Italiana Cittidina. Some traveling gypsies stole money in Palmer and were arrested here and were let go after returning the funds.

In case of war here, all children 0-5 age must register in case of any war emergency. The Memorial Day parade featured all of the women helping with the war effort, carrying flags, and all the Red Cross workers were in uniform. It was very inspiring. A meeting was held to help women bake without the use of wheat. Letter writers were encouraged not to send letters of discouragement to the men overseas. In August, families were rationed sugar to two pounds per month. People are substituting honey, corn syrup, and maple syrup. One meeting taught people about canning. People are now also using more eggs and there doesn't seem to be a shortage yet. The 4th Liberty Loan bonds went over the top and Negaunee raised its $325,000 share. The high school rally could not hold everyone. One nice write-up in the paper said, "No one knew how much happiness our young men put into our lives until they marched away."

Back in 1918 several people thought that the U.P. would be great sheep-raising country. 12,000 of them were shipped here, and some came to the south end of Marquette County. No further word ever seen in the paper. Workers hurt in the mines now get Workman's Compensation from an act passed in 1912. Several mines had programs subscribed to by all the miners and those monies are being returned now. Some miners will get up to $107.00. Autos are coming with cut-out mufflers, and many of the drivers are using them, getting attention for their loud noise. Some others on Iron Street are saying, "It's an outrage."

St. Paul's Parochial School has 322 children enrolled in September, and they graduated 21 from the eighth grade this past June. The Red Cross opened up a salvage shop with food and clothing and in the first month the profit was $1,245.00.

84. The First Fire Truck

On October 4, 1918, the newspaper notes that a fire at the Willman home on Teal Lake Avenue saw the first use of the city's "Motorized Fire Truck." It was noted that it had a satisfactory performance.

The 4th Liberty Loan Bonds sold here have been totaled up, and Negaunee went over the quota by over $50,000. The October Mine Report shows mine employment down by 500 over the previous year, and with 33 mines in operation. The Model Bakery in town has two new electric mixers, one for bread and one for cakes. Jaffet Rytknonen, Star Theater owner, lost his daughter, Tyne, age 15. In football, Negaunee beat Ishpeming for the first time in several years. The score was 3-0.

The city heard from the State Sanitary Engineer. They are to check that no sewage from homes on the south side of Teal Lake is going into the lake. At the Iron Herald, the owners have received a Line-o-graph, and the newspaper type will be set by machine for the first time rather than handset individual letters. With no alcohol use in town, Al Dyer and John Rock lost their jobs in law enforcement.

In November, Negaunee celebrated its going over the top with the bond purchases by closing down for a day. Mines and stores all closed, and a bank hung a sign up saying, "Everybody Celebrating. Come Tomorrow!" They didn't know it, but later in the same issue of the paper it was announced that the war was officially ended!

It was a bad autumn for the flu. Churches did not hold any Thanksgiving services, and the public was not allowed to attend school programs. Doctors are suggesting that you get an anti-flu vaccination. Only two cases here, but many in Gwinn, and schools closed in many towns. A week later in December there were 13 cases here. The theaters were closed until danger was gone. The library also closed. Precautions were encouraged: don't lick postage stamps, don't spend too much time doing business, and don't spit on floors and streets. Santa Claus was able to slip under the flu ban and visited children in the classrooms.

85. An Underground Cave-in

At the Negaunee Mine, there was another fall of a huge mass of iron ore in the number 2 shaft workings. Some noticed the ground in the area was "making weight," and warning went out to all the miners except two pairs of partners that were not reached in time. The wet and heavy ore entombed them when it came down and rescuers were quickly on the job. The next week's paper told of the survival of Wm. and Sylvester Arbelius, and Aino Eckoluoma. Mr. Wm. Medlyn had been buried underground.

You read about this store opening up in 1878. This ad is in 1971 Newspaper. Perhaps it was the longest existing store in Negaunee. It had a famous large kettle hanging in front of the store which is now the property of the Marquette Historical Museum. <u>Negaunee Iron Herald</u> advertisement.

AUCTION
SUNDAY OCT. 17th at 11 A.M.

535 W. Iron St., Negaunee. 90 year old Hardware Store known as Sawbridge Hardware. Complete stock to be sold.

— SPECIAL ITEMS —
Antique License Plates, Light from Old Ballroom, Dining Room Set, 2 Corner Booths from Restaurant, Window Seat high back, Radio-Phono Comb.; Antique Chest, excellent condition; 12 in. Band Saw, Double Holder Elec. Paint Mixer, Glass Cutting Frames, 5 Glass Showcases, Shelving, Old Catalogs and Magazines, Wooden Barrels and Hundreds of Other Old and New Items. Owner: A. Guizzetti.

1919

Already there is plenty of wheat available to buy and to eat, but warnings of not to waste it. Wm. Belstrom was the winner of a pony that Levine Bros. gave away for Christmas to the person with the most sales slips. Mr. Wm C. Sprout, who worked here in the 80's, has been now elected as the governor of Pennsylvania. The government in Washington has decided that "Near Beer" can be produced legally.

An important item, that could deserve its own fancy heading, is that women, for the first time, will be able to vote in the election here on April 7, 1919. They are reminded, however, to get registered now. Some passenger trains are ending service. The C&NW is no longer going to Calumet from Michigamme on the DSS&A rails, or from Ishpeming to Marquette. The CM&St.P may also drop its Champion to Marquette run on the DSS&A track.

In March, Negaunee won the U.P. State Title and headed downstate. Negaunee beat Flint, but lost to Holland. Holland eventually lost, also, and Cadillac was the State Title winner. Our local Board of Education invited Cadillac to come here and play Negaunee, and they did. We royally treated them, and the game was on a Saturday night at 9:00 p.m. when all the stores closed. Negaunee lost, 26 to 20, and hope to play them again some day. The quote in the paper was, "They are a fine bunch of fellows."

86. St. Paul's Church Burns

In April, a large fire was seen and fought by the town at the St. Paul's Roman Catholic Church. It was totally destroyed. Fr. Joseph Dittman expressed his thanks for the many kindnesses and for the high school teachers who helped to remove effects of the church and stored them at the high school. Luckily, the church is getting mining royalties from their sale of the old cemetery to the CCI, having had the mineral rights. This will be the basis of a new building fund. There was a problem of the original cornerstone disappearing and I was unable to find anything more about it. In any case, an architect engineer was soon hired and plans were made. It will be of solid brick. You can see it yet today.

The Methodists are concerned for their fellow Christians in Europe and locally raised the $12,290. that was asked for, plus $2,500. additional monies. Another main road is being tarred. This time the city is paving from here to the brownstone CCI shop buildings in Ishpeming. There is an ad for cherry pickers in the Door Peninsula of Wisconsin in June. The number of Negaunee graduates holds about the same every year. This year there are 37.

Jim Honka, the barber, has purchased an electric hair clipper. Clippers have been used for animals for some time, but are now perfected for barbershop use. There is an ad now for Wolverine Furnaces for central heating in homes. And the longest auto trip we have seen in the paper was when E. and H. Williams came here to visit their parents in their "little Ford" all the way from Portland, Oregon. It took them a month and two days, and said they had good weather all the way.

A war memorial is planned in the city square for those who died in the war. The war lasted 19 months, and 4,800,000 men served. Ten out of every 100 were National Guardsmen. In the physical exams, country boys did better than city boys. In August, there was a Grand Homecoming for the soldiers and sailors returning, with an automobile parade. Negaunee lost nine of her men and their names are listed. The celebration was for four days and all were told to wear a "poppy." The Zeba

Indians from L'Anse came to the homecoming and demonstrated their historic dancing and rituals for the public.

All drivers of vehicles and motorcycles now need to get a driver's license. Many are being stopped, and asking for licenses is getting quite common. A Negaunee branch of the Marquette County Historical Organization was organized, but their relationship to the Marquette group was never certain. As mines go, there are ups and downs. The Breitung Hematite is slowing down, but the Mary Charlotte recently hired more men.

There was, in October of 1919, a large fire in Humboldt which burned down much of the town, including the Pelmear home, their store, and the post office, all on the north side of the "South Shore" tracks (DSS&A). The railroad Station escaped destruction. The town now looked like a deserted village and never was rebuilt. The Mine Inspector's Report showed good employment with 3,265 men working underground and 1,783 on surface. The mines are fewer, with only 28 working. There were 16 deaths, all Finns and Italians, but 219 who were injured and lost more than 20 days of work. There were 426 injuries which resulted in a loss of 1 to 20 days of work.

The Sundberg Block is available for a city Community Center and could be renovated. It could come to a vote by the citizens. It was still a possibility several years later. There are still lots of wolves in Marquette County. In the first ten months there were bounties paid on 30 of them. The big concern as winter approached in December was that there is a shortage of coal. At this time, Negaunee gets all its power from the CCI and has shut down its own plant at the water building, and doesn't use any coal at all. We seldom read anything about the American Legion, but new officers were listed and the new commander will be Clarence J. Kearns, replacing Paul D. Barasa, who has retired.

Levine's employees had a very nice Christmas when they shared a Christmas gift of the profits the store made this year. The total was over $800 and one employee received $110.00. It was, said the newspaper, a "Family" type of Christmas. Mines closed down from Wednesday to Monday, and there was little social entertainment going on. There were no Christmas ads in the paper and no special Christmas edition.

This is a Michael Lempinen sculpture of an 1860 photograph of three men making a drill hole by hand. The title of the sculpture is "Double-Jacking." At the time of this photo in 2009, some Jackson Mine iron ore has not yet been placed at the base of the sculpture. It is located between the Jackson Mine pit one and the Heritage Trail, and next to the old Ishpeming-Negaunee Road.

Models for the sculpture, originally done in clay, were Mike, James, and Jon Reynolds, and Mr. Lempinen. It was fabricated by Mr. Rick Kauppila of U. P. Fabricating in Rock, Michigan.

1920

Mr. Louis L. Miller died. As a child, his mother had passed away, and then his father. He was the oldest of eight children, and they were all orphaned when he was eleven years old. As the newspaper put it, "Louis understood the duty of rustling for the support of his brothers and sisters." He secured odd jobs for a year or two and then Capt. Merry, the veteran manager of the Jackson Mine, admiring the lad's grit, gave him a job about the mine. There were no child labor laws to reckon with in those days, and the relatively large pay, which the youth received, went far toward establishing the family in comfort. He died at age 68.

A Negaunee girl, Miss Minnie Hakenjos, daughter of Wm. Hakenjos, has risen in the movie business out west quite quickly and now is the star in a new movie where she is called, Minnie Haken. Negaunee area farmers have gone to the Menominee Agricultural School for a meeting. A Ford service station will come to town. It will be in charge of assembling the autos that come in on the train cars and prepare them for sales. For Lincoln's birthday, the Mitchell Methodist Episcopal Church will have a special sermon.

The county wants to build a road from Marquette to L'Anse to connect with the road in Skanee without condemning too much private property. It would leave from the Negaunee area and go northwest. Ishpeming and Negaunee have gone together to hire a Community Service Director. He will emphasize scouting programs. Negaunee has also set up a new committee called the "Chamber of Commerce," to help in attracting tourists and advancing the town. Another tragic train death happened when two girls stood on the sides of the train track up against a snowbank made by the train plow, but could not get far enough back. The steps of the caboose swept them right under its wheels and both died terrible deaths.

The city is still thinking about an alternative water source. Now it may be from a spring-fed area. A week or two later they were again considering Lake Superior. The Italian Citizens League has 500 members in the area and is encouraging night school. Eagle Mills is growing again, now with many farms. They are seeking to have a post office restored there. Since saloons have "shut up shop," a third policeman has been discharged from the force as unneeded. A large quarry will be built behind Morgan Heights to mine and crush rock as "Greenstone," which will be shipped to a company making roofing.

The County Historical Society in Marquette is suggesting a large Historical Pageant. It will be on the south shore of the west end of the lake in a natural amphitheater setting. 3,000 people will be involved, and there were several rehearsals. On the day of the pageant it was estimated that half of the county was in attendance with special trains bringing people and all were asked to fill their automobiles. The weather was wonderful and a movie picture was made of the event.

The A & P (Atlantic and Pacific Tea Company) has come to Negaunee to be a distribution place for other area stores. Dr. Bell of Ishpeming has started horse racing at Union Park. Mr. Klein and his wife, in business here for over 20 years, are moving back to his home town of Muskegon. In September, because of a dry summer, the water levels are down and the CCI cannot make enough electricity to sell to Negaunee. Therefore, the city will restart up its light plant after 18 months of non-use. Diphtheria is quite prevalent again, but not alarming. Dentist Ben

Miller, born here and practiced here, committed suicide just before he was to be married. The city plans to install a new trunk sewer that will serve a city of 25,000. The high school will serve coffee and sandwiches to those students who wish the same, especially those from Palmer. The roller rink, now called the Adelphi again, will open for skating this winter.

The city purchased the most eastern lot on the north side of Iron Street, the Martell Lot, and will widen Pioneer Ave from Main to Iron St. In Sault, Ontario, the Algoma Steel Corp has shut down and over 2,000 people are unemployed.

This is a 2009 Photograph of the stairway up to the home at the corner of Cry and Snow Streets. The homes of the Jackson Location were moved in the 1950's and '60's. The Mather "B" mine would be undermining ore left at the Jackson Mine, and there was consideration of caving.

The Marquette and Western RR did not last long, but built this station between Gold and Silver Sts. It was originally further south in the present housing area.

1921

Jones and Laughlin Steel Co. has purchased the Breitung Hematite property. In order to save sewer system money, the city is thinking about dropping storm drains. After a large storm, however, it was decided that the entire system needs to be approved. Much water, not drained, caused much unwanted damage. The CCI has given the Boy Scouts some land for a scout camp to be built a mile north of Eagle Mills.

87. Negaunee's Big Ski Hill

In December of 1920, although late in the year, the city decided to build the largest ski hill in the U.P. on Calligan Hill. In February, the steel for the new ski scaffold arrived, but it will now be built on Teal Lake Hill where Pioneer Avenue has its terminus. Since the Ishpeming scaffold was blown down in the large 1920 summer storm, they will not be having a tournament this year, and came to Negaunee to assist in putting up its steel scaffold. It will rise 49 feet above the crest of the hill. No grading of the landing for the tournament will be done until next year. When February 22 arrived, there was not enough snow for the tournament and the new hill had to sit unused. In March, Negaunee boys were the basketball champions for the U.P. Members were Roy Broad, Charles Kangas, Rudolph Majhannu, Carter Curtis, Claire Knight, Gilbert Lindstrom, Theo. Sundquist, and Walter Collins. Negaunee lost downstate with two games to go.

Graduates, as usual, are in the 30 range. This year it was 31. A letter from Ohio was in the paper asking for someone to send her the recipe for that secret cake we have here, called saffron. In June, the body of the first Negaunee man who died in the war arrived here. His name was John H. Mitchell and the American Legion Post is named after him. The Lions' Club is very active now and their latest project is to find a swimming area for Negaunee. The first idea was to fill in a wall across the south (east) end of Teal Lake for swimming but keep it separate from the Lake. The Health Department turned it down. Many came to see a demonstration of a new modern gasoline powered fire truck with a water pump built on it and concluded it would be a good thing to buy one.

In August the Mary Charlotte hired 200 men after being closed for several months. However, by October the newspaper reports that local mining companies are unable to sell their iron ore. G. R. Granlund will open a Tire Vulcanizing Shop, to put good patches on tire inner tubes. I suppose we should put a picture of one in this book. The Chamber of Commerce has purchased the Sundberg Block and will maintain it. It is hoped that maybe some factories can be invited to use it.

An Athletic Association is being formed here to help promote at least the amateur sports. A report on night classes shows 21 boys and 24 girls enrolled and there are 38 studying citizenship and preparing for the same. Young offenders are being put on probation and must make headway in their schooling by advancing a grade. Since there are many men idle, the city in December asked all such men to register to see how great the problem was and 365 men came in to sign their names. A charity fund was started once again, and the first contributor of $500 was Mr. Wm. G. Mather with a personal check. Otherwise the town was quiet for the Holidays.

This was once the home of the Negaunee National Bank. It was temporarily used by the Negaunee State Bank after it burned down, and the First National Bank from 1973 to 1975.

1922

As Ishpeming is ready to have a ski tournament this year, Negaunee has given them back their Washington Birthday date and will hold theirs on February 11. The Negaunee basketball team is having a good year again and after ten games they are still undefeated. The Baptist Church at Grand Avenue and Case Street is having evangelistic services. It is tax time, and the citizens are hoping that the mines are not taxed too greatly or the taxes might kill the birds that lay the "golden eggs."

Last year, hardly any snow, this year in March, one with gale winds and eight-foot drifts. The city used their five-ton caterpillar tractor to pull their plow, but when they got to the end of a road, it was harder reopening the road and getting back.. Railroads had the same problems. When the Marquette basketball team finally arrived here, it was midnight, but a game was played. To fill in the time as the crowd waited, two junior teams played a game, and the city band gave a small concert. Negaunee won again. It was also good to have a storm because many men were hired to clear Iron Street.

The high school basketball team played an extra game with Stambaugh so that they could have both the Class A and B title of the U. P. They lost again downstate, this time to Kalamazoo, but in the losers' series, they took the state consolation cup. Also, in April, A. A. Alexander, of the Palmer Institute, is a new chiropractor in the Kirkwood Block. The Mitchell M. E. Church received a nice gift. Mr. Richard Stephens, a builder, died in California and left the church $5,000.00.

It was a real surprise this year to have a graduating class, the largest ever, of 49. The junior class currently is about 80, so the record may be broken again next year. The city made available to the citizens, $140,000. of sewer bonds and quite quickly, $40,000. of them were bought up. The balance was purchased by a Chicago firm. There is now a tourist map of the Trunk Line System of the State. It shows a road from Marquette to L'Anse via Skanee and the route that actually came to be via Three Lakes. Many of you have seen the large 7'3" statue of Abraham Lincoln at the old High School, or now at the Negaunee Historical Society. It was given to the school by the class of 1922.

It seems that the Negaunee State Police post will be moved to Marquette. In June, once again, the Teal Lake water is tasting "fishy." The reason is said to be that a recent storm stirred up the lake algae at the intake tub area. The city is now thinking of drilling some wells for water and storing it in a storage tank for use in busy usage times. As they cannot use sewer bond money for this, another $50,000 in bonds will be sold. Three wells will be drilled to see if the water is hard, or soft as lake water is. The result was that water was found, "but not a drop to drink." The wells encountered fine quicksand and it would wear out any water pump. Mr. Paul Honkavaara purchased the Muck property to the east of the fire hall. They will set up an electrical shop, especially for cars and their batteries.

Here is the Mine Report up to October. In 1920, 4,798 men were employed at the mines. Now it is only 2,276. The reason is that there are no shafts being sunk. Only the most experienced miners are still working, thus only one mine death last year. 118 serious injuries, however, and 221 slight injuries. The record year for mine employment was 1907 with 6,741 men. A favorite Negaunee teacher died, Miss Lydia E. Steel. She taught for 40 years and there are plans for a scholarship fund.

1923

The CCI gave its annual awards to the locations where their mines are. This year the awards went to the best kept premises, and the best vegetable gardens. Mayne loaders are now being used in drifts to pick up the ore or rock after a blast at the end of the drift. Rails are added and the loader moves into the rocks and pushes them into a bucket which in turn lifts up over the machine and into an empty car behind it. Seven are already in use at the Negaunee Mine, three at the Maas, and three at the Athens. Several are also in use in Ishpeming and Gwinn mines.

By March it still has been an easy winter. Mrs. Sam Mitchell died in Oregon. All of the CCI employees received a ten percent raise in pay. In June, Negaunee had 74 high school graduates, and St. Paul, 24 graduates from the eighth grade. Stores have decided to be closed every Wednesday afternoon until October 1. A large three-day homecoming is scheduled for the July 4th period. Mr. E. C. Anthony, during the homecoming, gave his story of the history of Negaunee since he came and then served in the Civil War.

The city received a notice from the Rolling Mill Mine that they did not own the land where the Palmer Road was and they wanted it moved as the ore under it will be mined. In July, the judge ruled in favor of the city in the road dispute, but suggested both parties get together and solve the problem. The mine decided it didn't need to be moved, but later the city did build a new road in that area. In August, it was brought to the attention of the town that the first car here was Mr. Wm. Maas's 1903 Oldsmobile. It is still running, but not used much. Mr. A. O. Sjoholm has been servicing it. There is also a 1903 Cadillac in town, owned by Bill's brother. It was ordered first, but arrived second in town.

The city is still trying to filtrate and improve the water. Still using its bond money, sand has been hauled in, and 340 feet of infiltration has been built. It is found, however, that it cannot filter enough water fast enough. The city will add 100 more feet of filtration sand. The Finnish Congregation in town has completed its new church at a cost of $26,000. Looks just like the one on US-41 now, and, of course, it is. A building that still exists in Republic is the downtown bank. In September of 1923, it was robbed of $10,200.

The Lions Club is very active and is still talking up a Community Center. Many families have left in the past year, and school enrollment is down, and they feel that not only does the town need more jobs, but better community activities. If you have read this book from the beginning, you have also read 50 years of the Iron Herald. They celebrated their anniversary in November. The post office is on the move again, and is now on the first floor of the MacDonald Opera House. At Christmas, the high schoolers personally played "Santa Claus" to 14 needy families.

This is believed to be Streetcar # 15. It was reclaimed in 2007 as a Midway Cabin on old US-41. It now belongs to the Negaunee Historical Museum and will be restored.

1924

Bad news began the year for the city regarding the sand water infiltration project. Another 300 more feet of sand will be needed. There was bad news also for the Methodists when a bad fire caused much damage to their church.

In March, as is often usual, the Negaunee basketball team won the U.P. Championship and the trophy cup. Then in April there was another large fire, this time at the Negaunee Hospital on Teal Lake Avenue and three patients died from flames and smoke. In the spring, more and more radios were becoming popular and about 50 people now are busy finding distant radio stations, with people and music reaching their ears during the nights..

In May, this famous action took place. It was the planting of 25 trees eastward towards Negaunee from the Ishpeming boundary line. Each tree had a plaque, recognizing every local man who died in the World War. Plaques are now at the Negaunee History Museum, following the removal of the old, grown trees.

The economy is not good and the CCI mines report, that to keep men working, and to avoid a complete shutdown later, their mines will all go to four-day weeks. The businessmen in town would like to see more Michigan highway traffic (No U.S. highways until 1932) coming down Iron Street, so they petitioned the state to have M-15 and M-35 traffic come from Main, to Pioneer, to Iron, and down to Silver, rather than using Jackson St. They suggested further, that the road skip Silver and keep going west on Iron, up past the Jackson Mine monument and on to Ishpeming. On September 5, it snowed quite heavily at the Water Works for about five minutes in the morning. The state replied, regarding change of roads and said they will remain the same, using Main to Jackson, to Silver to Ishpeming. This caused a problem for Standard Oil and their plans for a station by the Fire Hall. They are now seeking a lot by Teal Lake and Main, but cannot find anyone who wants to sell. Negaunee women are organizing a rally to get women out to vote and use their new right.

We now return in November of 1924 to the disappointment that began the year, and that information is that after all the year's work, the water filter flume did not still pass the test. The sand is not of good quality, and the report shows that it could never be completely successful. The lake is also getting lower for some reason. When December comes, there are two nice items in the paper. CCI employees welcome the news that their four-day schedule will return to a full-time five days a week. The second news is that the Women's Club delivered Christmas baskets to 70 homes.

This is a 2009 photo of a tunnel area that can be walked to on the south side of the old highway at Pit No. 2. This is a dry pit (down to the right) and is not fenced off. There is a concrete cover over an old shaft in it.

1925

We always seem to talk about how much snow we used to get years ago. However, many of the winters seem to be much easier than at present. Again in 1925, at the end of January, the paper reports that the city now has three brand-new large plows and tractors to push them, but there is not enough snow for them to be tried out. Negaunee's school colors are blue and gold, but the people by March are only blue. After three U. P. Championships in basketball, this year they lost it. They tried to find a way to go downstate anyway to show their talent, but it was ruled illegal.

The news was soon forgotten no doubt, when in the same newspaper, the CCI announced that it had received a contract to ship 2,600,000 tons of ore in the next several years, at a half million tons a year from the Negaunee, Maas, and Athens mines. At Teal Lake, when the ice went off in April, the lake was down eight feet from its high in 1893. The city is thinking of getting water from Morgan Creek in Section 27 of R48, T26. In May, Mr. E. R. Tauch of Marquette decided to buy the CCI Greenhouse on Gold Street, but after a year, he sold it back, but stayed on as the manager.

In July there was a bad rain and hailstorm with much damage, the worst in many years. The city now has the Teal Lake Sports Field, and will build a modern 1000 foot-long stadium. On the next ballot will be some choices for the people regarding a Community Building. The first vote is to allow the indebtedness. If this passes, then one of two other choices will be completed. One is to buy the Sundberg Block and the second is to build a new building.

The new road that was built by the city for the Rolling Mill Mine in 1925 is now having caving problems. The city was also checking to see if the water from the Maas Mine could be run into Teal Lake to solve the water level, but it is contaminated. In September, the bond proposal for a Community Center was turned down and seemed to be the end of this story after eight years of work. However, the very next month, the Chamber of Commerce, the Common Council of the City, and the Lions Club are trying to work out a plan to utilize the Sundberg Block.

The City Water Commission is now thinking that part of the water level problem is the old wooden pipes which could be replaced with steel ones along Teal Lake Avenue into town. They also believe that if the city used all water meters, the use of water would go down about a third. In the middle of October there was a large snowstorm. Phones were out, there was no electricity, Street cars were inoperable, but sidewalks got plowed using a double team of horses.

No real Christmas news except that we see that things are also tightening at the post office. For the first time ever, they will not be open on Christmas Day, and no one will work any overtime.

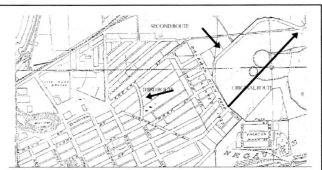

Before U.S.-41 came in 1932 from Negaunee to Marquette along Teal Lake, there were three routes. The NE arrow is the original route of Main Street going past the original cemeteries. The second route went to old Baldwin-Kiln Rd., then NE on a new route to the road at the cemeteries. Route 3 (left arrow) was formed as caving to the east continued. From the Negaunee Historical Museum map collection.

1926

Mining reports in the paper conclude that it is far less likely of finding ore in the area in the future than it was in the past, and that ore will be much more expensive to mine. An even worse report in January was that at a service of the Methodist Episcopal Church, three hooded Klu Klux Klansmen came down the aisle during the closing hymn and handed an envelope of money to the pastor, and then left. Nothing further ever found in the paper.

A lot of original townsmen have been passing on, and in January there was the death of Antoine Deloria. He was here as the foreman of several teams that helped to take the ore from the Cleveland and Jackson Mines over the Plank Road to Marquette. "Until comparatively recently, when the advent of the automobiles demanded complete rebuilding of our highways, evidence of this old plank road were still discernible in the highway between Negaunee and Marquette. He also cleared the land for a railroad that was never built from Marquette southwestward to the line of the C&NW not far from Cascade."

It was Holy Week at the churches in March, and St. Paul's and the public schools both closed for Maundy Thursday and Good Friday. The city is seeking to get an army cannon for the city hall square. The government is giving them out for parks. The Street Railway gives its first hint that it may stop running, or at least stop in the summer time when cars are in use. They say that they still might compete in the winter when they run better than busses.

Standard Oil is still looking for a gas station along the traveled highway by Teal Lake and Main Street. Negaunee has 72 graduates, 36 of each sex. St. Paul's had 23 eighth graders

graduate. Prohibition is on, and a huge moonshine still was found by Federal officers. Frank E. Roberts had 25 gallons of liquor and 1000 gallons of moonshine mash in barrels. Mr. Rytkanen had a contest for a name of his new theater. You know the winning name. It became the "Vista." The Star Theater came to an end and is being remodeled into an auto showroom.

Because of the bad economy in the past few years, there have been few large carnivals or circuses to report on. This summer, however, the Wortham Entertainment Shows arrived in 34 double railroad cars and were sponsored by the Sons of St. George. The Chautauqua also came to the outskirts of town, and provided sports, music and educational lectures for several days, using tents. There was a good attendance, but this year the loss was quite great and another fundraiser will be held to pay off the debt. In Palmer, the Empire Mine has been purchased and a new crusher is already being put in place for mining. It has a siliceous ore which is in demand.

The newspaper notes that the final stretch of sidewalk under the LS&I viaduct at Teal Lake Avenue is not completed, and that people should note that they should not step into the roadway when walking under it. There have been several close accidents. More word about the Klu Klux Klan. They have a huge tent outside of town for Labor Day. Even though the weather was bad, many people went to visit it. There are also many auto accidents now. In one week's paper alone, a Gwinn woman died, and an Ishpeming man, and there was injury of a score of persons with some wounds of a serious nature.

The Chamber of Commerce announces that they have arranged for a glove factory to come in the Sundberg Block. The next week, the city received a letter from the mines saying they are against the city putting money into a private company and so everything stopped. The third week we learn that the building is being remodeled for a glove factory anyway. (The judge threw out the mine injunction.) A new furnace was also installed. A month later, the glove machinery was on its way to Negaunee, and soon installed. The paper announced on December 3rd that, "We are turning out Negaunee Gloves." With a change of heart, the local mine officials reported that they will ask their headquarters to pay for half of the $25,000 Chamber of Commerce expenses incurred in obtaining the glove factory.

Considering there is little mining news, Mr. Mather arrived from Cleveland and explained to the miners and public that the old Bessemer furnaces used native iron ore from here. Now he reports that the new furnaces in the cities are large, open hearth ones that use both pig iron and scrap steel, so there is less need for native ore. He also says that steel will always be needed, and more products will be needed in the future in both Europe and the USA.

The large Barnes-Hecker mine disaster occurred in November, west of Ishpeming, when a large part of a very wet area poured down into the mine and only one man survived. Fifty-one died. The mine was closed forever. Mr. E. C. Anthony died in December. The CCI is buying houses on the east side, and recently bought the Tom Pellow home as they are leaving for California. Local fishermen are happy they can boat navigable streams and that private land owners only own the land up to the water.

1927

January and February papers have no news of any note, and several of the papers are so full of ink that they are unreadable. By March we find that the High School boy's basketball team is in true form and have lost only one game. There is some talk of a municipal pool, and every week there is now a single-frame newspaper cartoon. Several are about women's skirts and dresses. They are now up to the knees. Mr. James Jopling, a CCI engineer reminisced in the Iron Herald about the early day mining town of Negaunee. He said that the town was never wild as the people here were more religious and as mining was also opening in Montana at the same time, most of the wilder miners went west.

There was bad news for the basketball team. The Blue and Gold is out of the tournament, the first time since the U. P. tournament was inaugurated, and the "gloom is correspondingly great." Details are in the March 11, 1927, paper. Intramural basketball has begun for 300 boys in grades seven through 12 at the gym, all day on Saturdays. In April Mr. K. I. Sawyer, the Marquette County Road Superintendent, announced that the new US-41 route from Florida to the Copper Country will have Negaunee on the route. Albert Larson and William Conners were both killed in car accidents in April.

Times have changed the train schedules. Trains formerly ran more in the summer and less in the winter. Now they will run less in the summer and more in the winter ~ automobiles are the reason. The Negaunee Police Chief is now Bert Agnoli. It looks like the town might grow. In July there were only 3 deaths and 15 births. For the year to date in August there were only 42 deaths, but 100 births.

88. Moving East Side Homes

The August 12, 1927, newspaper had a very important notice from the CCI, both in an article and an advertisement: "The CCI has made a start upon the plans for clearing the residential area to the eastward of the old cemetery, preparatory to the mining operations which will eventually be inaugurated. A contract has been let to a Hibbing firm for the removal of 25 homes to a new location east of the Collins addition. Concrete foundations for the structures are already in place. It is expected that active work will begin next week and the plan is to move "a house a day."

89. Streetcar Service Ends

On the evening of August 20, 1927, the Michigan Gas and Electric Company had their Ishpeming-Negaunee streetcars make their final trip before beginning to take up the rails and ties. The Twin City Motor Bus Service duplicated the routes and the time schedules starting the next day, and placed an ad in the paper for the same. The Gas and Electric Company noted that the busses, with continual road improvements, should do as well as the streetcars did in the winter time.

Mr. John Mayne, the Negaunee man who built a mine that is electrified into a working model, will have his work shown at the Michigan State Fair in Detroit. John spent all of his life in the mines here and has exhibited his creation all over the U.P. and in Wisconsin and in Minnesota. The Negaunee Hospital was purchased by the CCI as a mine hospital. Doctors Andrus and Sheldon wish to retire. At the end of September, however, Dr. Andrus opened a private office in the former Star Theater Building. There are going to be more activities for girls in town now that the Camp Fire Girls have formed groups. The city will also build a bandstand for the City Band. It was undecided at first whether it should be in the City Square, or on the east side of the City Hall.

The Queen Mine still has over $9,000 in funds given by miners as a insurance for themselves. The state wants the mine to try one more time to find the miners before the money is turned over to them. The last interesting news was in December when Henry Ford announced that he will replace his "Model T" with a new model that has many of the features of more expensive automobiles.

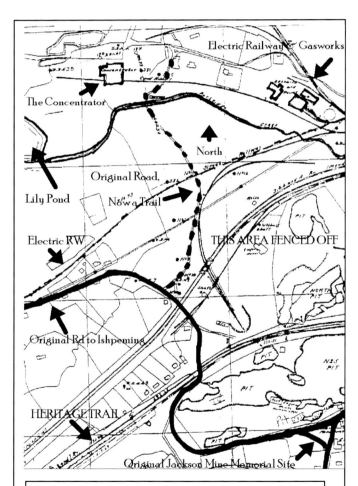

Here is a map of the western area of Negaunee and the Jackson Mine as it was a hundred years ago. It shows where the Concentrator and the Electric Railway barns, and the Gas Manufacturer Plants were located. Negaunee Museum Map; edited by the author.

1928

One thing is lacking at the cemetery says the Women's Club. It is a real chapel at the cemetery for mourners and others during the winter months. They have approached the city about the matter. Dr. Burke sees the need for a new city hospital, and has purchased the residence of the Crane's on Cyr Street. It will be remodeled into a modern institution by March.

With automobiles, the city now has its tourist park and it is very popular. Some say it is the best in the northwest. Many salespeople stay here for long periods, going back and forth to other areas of the U.P. A study of Jackson Park showed that in 1927, 427 camping parties used the facilities, including water, showers, and baths. The majority of campers stayed from four to six nights. (Many townspeople from the surrounding areas also paid money to use the showers.)

There are many local shows, as well as out of town acts and plays taking place, but no longer at the MacDonald's Opera House. They all seem now to be at the new Vista Theater. Plus many good movies are being shown on many nights. In March, the Reverend James Stanaway died. He had started Sunday Schools in many small towns all over the U.P, and northern Wisconsin. He was born in Cornwall in 1857. Two C&NW freight trains hit headon in the Goose Lake section of track. Both ends of the train kept piling up after the engines hit each other.

Time is moving on and now all the cities are starting to build airports. Ishpeming will put one at Union Field, says Negaunee. There are now thoughts of being able to fly to Chicago. Hancock and Menominee are also building already. Another accident happened at the S Curve west of Morgan Heights with two girls dying. It was named, "Dead Man's Curve." Mr. and Mrs. Mitchell of Champion will build a creamery here in the Torreano Building. Farmers are looking forward to bringing their milk and cream there.

90. Abandonment of Streets

With the Maas, Lonstorf, and Mitchell Addition already abandoned for the Negaunee Mine, the city was approached (May 11, 1928, newspaper) regarding the first major "vacation" of downtown Negaunee. There was both a published notice in the paper, and a long article for the abandonment of part of the Harris Addition, and a rearrangement of the northeast portion of the city. Before actual abandonment of several additions, the CCI proposed a new addition which they would build at no cost to the city. It would consist of 155 lots on a triangle of land north of Cherry Street. There will be a street north and parallel to Cherry. The present Baldwin Kiln Road will be abandoned, and the new Baldwin Kiln Road will be the west end of the new triangle addition. The new alleys will be wider to accommodate snowplows so that automobiles can get out of their garages.

91. Ideas for an Airport

Negaunee began to investigate for a airport in June of 1928. They believe that even the mail may travel by air eventually. Two weeks later the city found the old F. W. Read farm, east town and just on the west side of the old Eagle Mills. It is the top of a large hill. The landing strip will be 2000 feet long and 150 feet wide. The newspaper gives the access routes. From the north, go south from the "concrete highway" (a surprise), or turn left from 492 and go up and around the cemetery and then go north and hang to the right. Or one can start at the bottom of the hill at Eagle Mills and walk up about 340 yards. Two weeks later, there was already an official opening of the airport, and some local men have purchased a large plane that would hold at least 15 people. They gave trips up in the Stinson-Detroiter all day, but on what became the last trip with six people aboard, the plane hit an air pocket and dropped into the cleared brush, 15 feet short of the runway. Passengers were shaken up and had some cuts and bruises, but all were safe. The plane was later sold by local men James A. Thomas, and Albin Belistrom to U.P. Airways after repairing the plane.

Now we will return to the vacating of more city additions. In the paper of August 3, notice was given of a petition filed in court by the CCI for the abandonment of the plat of Corbit's (Corbet also) Addition, No. 2, certain other plats, the Baldwin Kiln Road, and the closing off of several streets at their east entrances. The reason is that the mining of the Maas Mine under those areas may cause caving. The mine notes that they have already bought most of the individual pieces of property. The road designed a few years ago when Main Street was abandoned to go around the abandoned cemetery area will now itself be abandoned and another new road will be made to go to Marquette. The company notes that the added mileage will be less than a block.

A meeting was held with the public invited. Three area men spoke up regarding having to go west on their street to now get the road east to Marquette. People also felt that their homes, being left, would go down in evaluation, living on dead-end streets. The meeting stopped at 11:45 p.m. and was rescheduled. The paper noted: "Clearly Negaunee cannot afford to impose hardship on the organizations upon which it is dependent for support."

Two weeks later the newspaper printed a map of the area to be abandoned, and just before the next meeting. The company has not been able to close off streets, but is able to be building the new roads. The new route will be to use Cherry Street and extend it east to the north end of the old cemeteries. The August 24th paper received a report from the Cleveland Cliffs Iron Company that they have bought most of the homes and are offering good settlements. The townspeople again spoke as well, saying that as citizens they had paid good taxes to build all these streets with water, sewer, and city electricity, and the mine was not reimbursing the town at all for any of this. It was also suggested that the mine take all the ore from other areas not under homes first. Some also had the insight to note that "this would be just the first bite off of the city streets." Several people did not want to sell, and others who sold said they were happy with what they received. It was then time to vote. One councilman was absent, and the vote was a tie. There will be another meeting.

CCI is so anxious to begin mining under the town, that the next week the newspaper noted that 20 Maas Mine men had been let go for now until the mining of that area can begin. On October 5 the city and the CCI reached a street agreement. Healy Avenue will be extended to the north, but swerved to the east slightly from Mitchell to Cherry, and become the feeder for the altered M-15. The Mining Co. also noted that they will buy property in the Gaffney's, the Kirkwood's and the Kellan's Additions. If passed, the mine will immediately put up a barrier on Main Street at the SW corner of Corbit's #2 Additon, and at Case, Race, and Pine Streets where they join Grande Avenue. All of this took place and was solved after three days in court.

In September there was the very first U.P. State Fair in Escanaba. It was noted that the distance did not seem to have any effect on local attendance, and that it was a success. I made one other note as I read through the balance of 1928. After many deaths of the town's leading men in the decade of the 1920's, the wives are now passing on, with several in each weekly newspaper. One had been married twice and had lost her second husband 25 years ago. She was 92.

REALTY TRANSACTIONS 1929
IN "EAST END" TO DATE

The following are the residence properties in the eastern part of the city upon which the Cleveland-Cliffs Iron Company has exercised the options it held for their purchase:

—Mary C. Gaffney Addition—

Main Street—	Lot	Block
J. J. Baldo	14	1
Case Street—		
Everett A. Annelin	10	1
Anders Olverson	7	2
Henry Henrikson	10	2
Emma Dawe	11	2
Charles Waasberg	13	2
Julius Lahti	14	2
Sarah Rund	2	2
—Kirkwood & Kellan Addition—		
Park Street—		
Joseph E. Roberts	7	1
Gustav Dahlstrom	8	1
Albert Sundin	11	1
Charles Larson	13	1
Jane Chapman, et al.	16	1
William H. Holman	16	1
William Sundquist	6	2
Christian Wiik	7	2
Raatikainen (1/4 interest)	11	2
Charles Augustson	14	2
William J Stephens	15	2
Ruby Trathen, et al.	19	2
—Maas, Lemster & Mitchell Add.—		
P. Rasmus Christensen, 13-14	2	6
John Tuuri	1-2	6
A. William Koski, et al.	1-2	7
Otto Lindstrom	7	7

1929

The January 11th newspaper told us that up to that date, there has been little snow, even for the Christmas holidays. The Standard Oil Company had a happy New Year. The Goodman Family sold their home to Standard Oil for a gas station at the corner of Teal Lake Ave. and Main Street. A happy New Year for the public also as the Vista Theater will begin showing its first "Talking Movies." A Mary Pickford film will be among those shown. The City Band is preparing for its 50th Anniversary at its new bandstand at Case and Kanter. They began in 1880.

Palmer had 14 graduates, and Negaunee had 77. There was a nice article about the Ishpeming Theater and Negaunee people had a lot of nice memories. Iron Street will get its first repaving since the original tar some years ago. It has held up well until the removal of the Street Railway tracks. The city has begun a new search for city water and again will drill some new test wells. The Sundberg Block with the Glove Company has been paid for and the Chamber of Commerce would like to develop recreation for the city at Teal Lake. There are still horses in use, and the one at the Breitung Hotel that went to the train station has now, like many others, been retired.

Early in the year, the city ran into a legal problem of extending Healy Avenue to Cherry for the new road to Marquette. The Kellan Estate owned a lot, #1 of the Kirkwood-Kellan Addition, and some of the whereabouts of the owners have not been located. They would like to offer $4,000. Then it was reduced to $3,000, and the CCI said they would only pay the city $1,000 for it. The city went ahead with the condemnation proceedings against the Kellan Estate, and a jury was selected. However, the

Kellan Estate settled for the $3,000. and the CCI roadwork continued.

The DSS&A, known in the paper now as mostly the "South Shore," is going to drop several train runs for passenger service, many of them being short runs between towns. The name "NEGAUNEE" has been placed in large letters on the roof of the City Hall so that it will aid aeroplanes. People were excited to see a squadron of 19 Army planes fly over in the winter from a Lower Michigan base on their way to Spokane, Washington. City records for the year of 1929 showed that 59 boys were born, and 58 girls, and that there were slightly less deaths than the 117 births. It was also a very good year for mining and families. Escanaba shipped over 800,000 tons more of ore in the 1929 season over the previous one, and for the Marquette port, it was almost a million more tons. Things again are looking up. However, the Great Depression is about to begin. I hope you have enjoyed The City Built at the Shiny Mountain. Thanks to the Iron Herald, "The only paper that gives a darn about Negaunee."

Here is a map of the Negaunee and Palmer Mines. Several Mines are not shown. Mines also changed names from time to time. The small a, b, c. mines are represented in the "Regent Group." The road going east is now 480, going west and north-east is US-41 (top) and M-28 (middle)

The bottom road going east-west went from Palmer to the Cliff's Drive and is now gone. The black dot was the home of the author's wife. Map is by the Lake Sup. Iron Ore Assoc. in October, 1937.

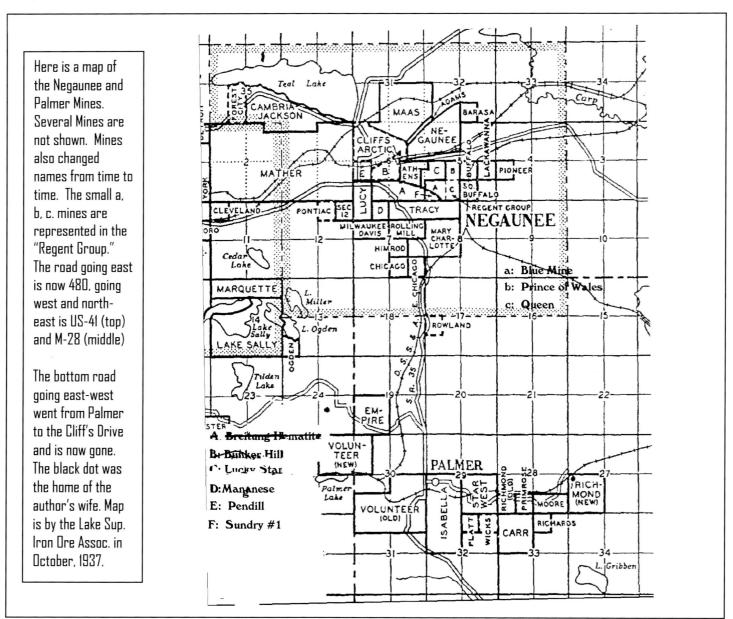

a: Blue Mine
b: Prince of Wales
c: Queen

A. Breitung Hematite
B. Bunker Hill
C. Lucky Star
D: Manganese
E: Pendill
F: Sundry #1